VEDIC
MATHEMATICS
FOR STUDENTS

notionpress
.com

VEDIC
MATHEMATICS
FOR STUDENTS

LEVEL – 2 OF 5 SERIES

NAVA VISION

Notion Press

Old No. 38, New No. 6
McNichols Road, Chetpet
Chennai - 600 031

First Published by Notion Press 2018
Copyright © Nava Vision 2018
All Rights Reserved.

ISBN 978-1-948424-22-6

DEDICATION

India is land with great history and birth place for Vedas. It has been home for many great minds like Aryabhatta, Bhaskaracharya, Bramhagupta. We want to dedicate this book to all the great ancient and modern Indian Mathematicians who make India proud for the centuries to come.

We would like to thank "Jagadguru **Shankaracharya Bharathi Krishna Tirthaji Maharaja,**" the father of Vedic Mathematics for his contributions to the field of Mathematics.

Swami Bharathi Krishna Tirthaji

Producing this book is indeed a group effort of our Nava Vision Team to help getting the valuable and unique content, binding up expertly steered the project. We would like to thank our whole team of NavaVision for their immense effort put together that lead to publishing of these **Five level books** on Vedic Mathematics.

CONTENTS

CHAPTER 1

- Urdhva Tiryagbhyam

  - ❖ Urdhva Tiryagbhyam for 2-Digit Numbers without Carry

- Urdhva Tiryagbhyam for 3-Digit Numbers without Carry

- Learn Tables in Simple Way! – Trick for Table 7

URDHVA TIRYAGBHYAM

Meaning: Vertically and Crosswise

This Sutra is used in all cases of **multiplication** which uses the concept of multiplying vertically and cross wise. This method is simpler and easier to understand than the conventional method; and computation time is less. With much practice, this can even be done through mental calculations.

This has already been discussed in Level -1. This section has a recap of the same to learn more advanced concepts.

URDHVA TIRYAGBHYAM FOR 2-DIGIT NUMBERS WITHOUT CARRY

We see here how to multiply any 2-digit numbers in general. Note that in the below examples there is no 'carry'.

Let us consider an example, 11 × 12

Step 1: Multiply the digits in unit's place i.e., $1 \times 2 = 2$

$$
\begin{array}{rr}
1 & 1 \\
\times \quad 1 & 2 \\
\hline
& 2 \\
\end{array}
$$

Step 2: Multiply the digits cross wise and add both products i.e.,

$(1 \times 2) + (1 \times 1) = 2 + 1 = 3$

$$
\begin{array}{rr}
1 & 1 \\
\times \quad 1 & 2 \\
\hline
3 & 2 \\
\end{array}
$$

Step 3: Multiply the digits in ten's place i.e., $1 \times 1 = 1$

$$
\begin{array}{rr}
1 & 1 \\
\times \quad 1 & 2 \\
\hline
1 \quad 3 & 2 \\
\end{array}
$$

Result: 11 x 12 = 132.

All the 3 steps can be represented in a single step as follows,

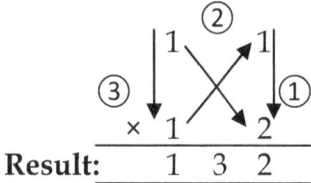

Result: 1   3   2

Let us consider another example, 13 × 12

Step 1: Multiply the digits in unit's place i.e., $3 \times 2 = 6$

$$
\begin{array}{ccc}
 & 1 & 3 \\
\times & 1 & 2 \\
\hline
 & & 6 \\
\hline
\end{array}
$$

Step 2: Multiply the digits cross wise and add both products i.e.,

$(1 \times 2) + (1 \times 3) = 2 + 3 = 5$

$$
\begin{array}{ccc}
 & 1 & 3 \\
\times & 1 & 2 \\
\hline
 & 5 & 6 \\
\hline
\end{array}
$$

Step 3: Multiply the digits in ten's place i.e., $1 \times 1 = 1$

$$
\begin{array}{ccc}
 & 1 & 3 \\
\times & 1 & 2 \\
\hline
1 & 5 & 6 \\
\hline
\end{array}
$$

Result: 13 x 12 = 156.

All the 3 steps can be represented in a single step as follows,

Result: 1   5   6

URDHVA TIRYAGBHYAM FOR 3-DIGIT NUMBERS WITHOUT CARRY

Multiplication of 3-digit numbers follows the same procedure as multiplication of 2-digit numbers. We'll learn more with the below examples.

Let us consider an example, 231 × 302

Step 1: Multiply the digits in unit's place

i.e.; $1 \times 2 = 2$

$$
\begin{array}{cccc}
 & 2 & 3 & 1 \\
\times & 3 & 0 & 2 \\
\hline
 & & & 2 \\
\hline
\end{array}
$$

Step 2: Consider only the unit's and ten's place digits. Multiply the digits cross wise and add the products

$(3 \times 2) + (1 \times 0) = 6$

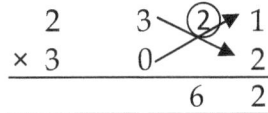

Step 3: Consider all the digits. Cross multiply and add the products.

$(2 \times 2) + (1 \times 3) + (3 \times 0) = 4 + 3 + 0 = 7$

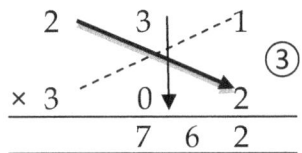

Step 4: Consider only the ten's and hundred's place digits. Multiply the digits cross wise and add the products

$(2 \times 0) + (3 \times 3) = 0 + 9 = 9$

$$\begin{array}{cccc} & 2 & 3 & 1 \\ ④\times 3 & & 0 & 2 \\ \hline & 9 & 7 & 6 & 2 \end{array}$$

Step 5: Multiply the digits in hundred's place

$(2 \times 3) = 6$

$$\begin{array}{ccccc} & 2 & 3 & 1 \\ ⑤\times 3 & & 0 & 2 \\ \hline \textbf{Result:} & 6 & 9 & 7 & 6 & 2 \end{array}$$

Result: 231 x 302 = 69762.

Let us consider another example, 141 × 212

Step 1: Multiply the digits in unit's place

$1 \times 2 = 2$

$$\begin{array}{cccc} 1 & 4 & 1 \\ \times 2 & 1 & 2 & ① \\ \hline & & 2 \end{array}$$

Step 2: Consider only the unit's and ten's place digits. Multiply the digits cross wise and add the products

$(4 \times 2) + (1 \times 1) = 8 + 1 = 9$

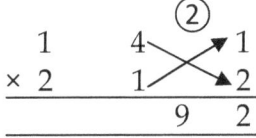

Step 3: Consider all the digits. Cross multiply and add the products.

$(1 \times 2) + (2 \times 1) + (4 \times 1) = 2 + 2 + 4 = 8$

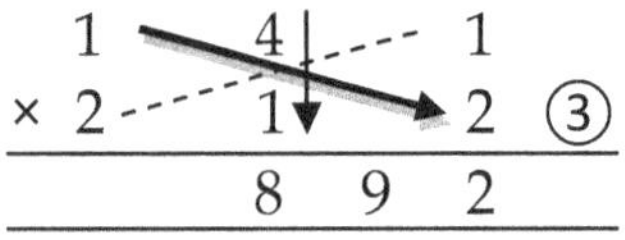

$$
\begin{array}{ccc}
1 & 4 & 1 \\
\times \ 2 & 1 & 2 \ \ ③ \\
\hline
8 & 9 & 2
\end{array}
$$

Step 4: Consider only the ten's and hundred's place digits. Multiply the digits cross wise and add the products

$(1 \times 1) + (4 \times 2) = 1 + 8$

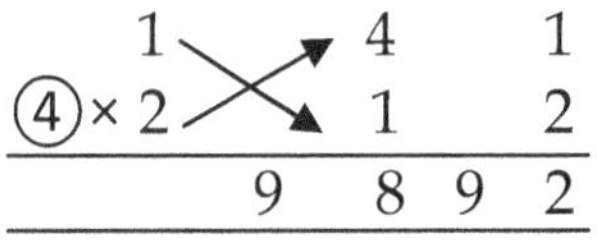

$$
\begin{array}{cccc}
 & 1 & 4 & 1 \\
④ \times 2 & & 1 & 2 \\
\hline
9 & 8 & 9 & 2
\end{array}
$$

Step 5: Multiply the digits in hundred's place

$(1 \times 2) = 2$

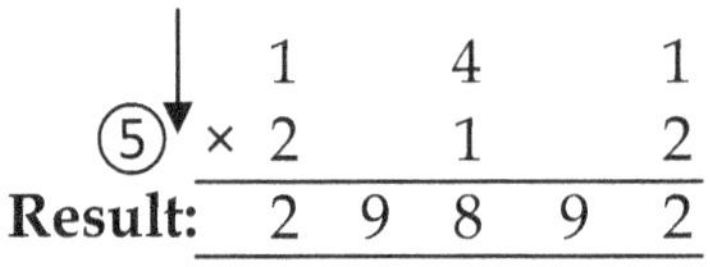

$$
\begin{array}{cccc}
 & 1 & 4 & 1 \\
⑤ \ \times \ 2 & & 1 & 2 \\
\hline
\textbf{Result:} \quad 2 & 9 \quad 8 & 9 & 2
\end{array}
$$

Result: 141 x 212 = 29892.

LEARN TABLES IN SIMPLE WAY!

TRICK OF MULTIPLICATION TABLE FOR 7

To form the multiplication table for 7, there is a trick that can be followed instead of memorizing it. To get the trick, form a grid with 9 cells as shown below and write then write the numbers in the order, as shown below.

0	1	2
2	3	4
4	5	6

7	4	1
8	5	2
9	6	3

Now club both tables together to get 7 table as shown in the figure on left.

So, the complete table can be represented as shown in the figure on right.

07	14	21
28	35	42
49	56	63

7 x 1 = 07	7 x 2 = 14	7 x 3 = 21
7 x 4 = 28	7 x 5 = 35	7 x 6 = 42
7 x 7 = 49	7 x 8 = 56	7 x 9 = 63

EXERCISE 1.1

I.   **Multiply the below numbers using Urdhva Tiryagbhyam method**

a. 11×15

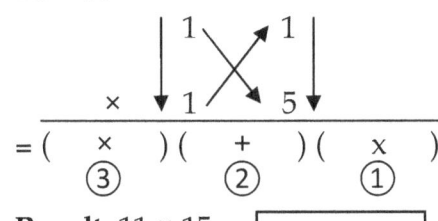

= (×   ) (+   ) (×   )
    ③   ②   ①

Result: $11 \times 15 = $ ☐

b. 23×12

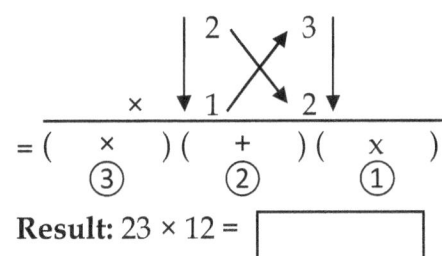

= (×   ) (+   ) (×   )
    ③   ②   ①

Result: $23 \times 12 = $ ☐

c. 10×11

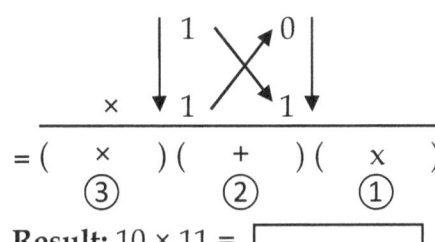

= (×   ) (+   ) (×   )
    ③   ②   ①

Result: $10 \times 11 = $ ☐

d. 10×24

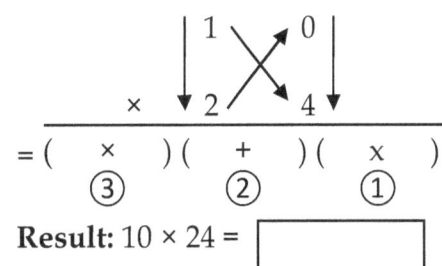

= (×   ) (+   ) (×   )
    ③   ②   ①

Result: $10 \times 24 = $ ☐

e. 11×33

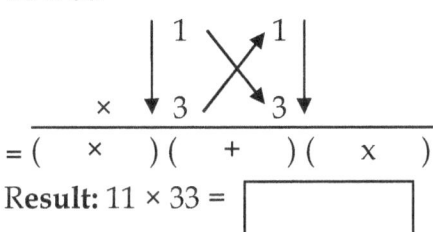

= (×   ) (+   ) (×   )

Result: $11 \times 33 = $ ☐

f. 11×22

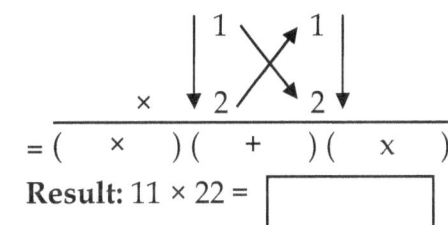

= (×   ) (+   ) (×   )

Result: $11 \times 22 = $ ☐

g. 12×33

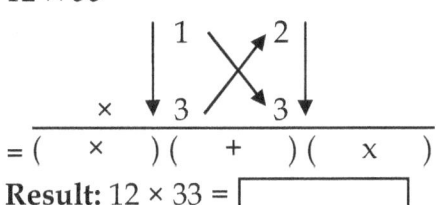

= (×   ) (+   ) (×   )

Result: $12 \times 33 = $ ☐

h. 23×21

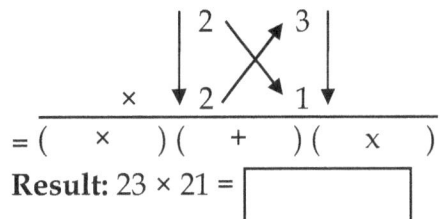

= (×   ) (+   ) (×   )

Result: $23 \times 21 = $ ☐

II.   **Multiply the below numbers using Urdhva Tiryagbhyam method**

a. 23×11

b. 11×44

c. 12×14

d. 21×14

e. 31×22

EXERCISE 1.2

I. Multiply the below numbers using Urdhva Tiryagbhyam method

a. 131×201

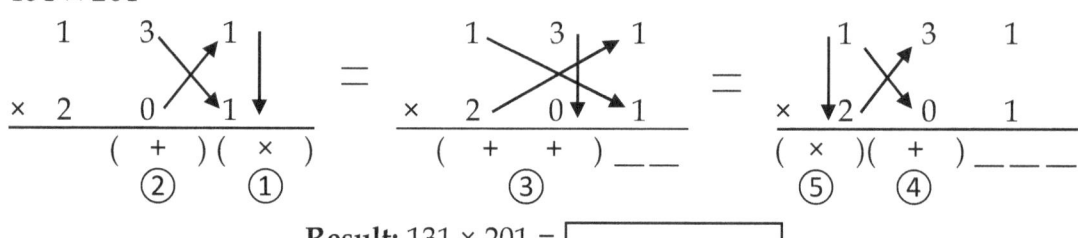

Result: $131 \times 201 =$ ⬚

b. 123×211

Result: $123 \times 211 =$ ⬚

c. 301×112

Result: $301 \times 112 =$ ⬚

d. 231×121

Result: $231 \times 121 =$ ⬚

e. 301×212

Result: $301 \times 212 =$ ⬚

f. 131×220

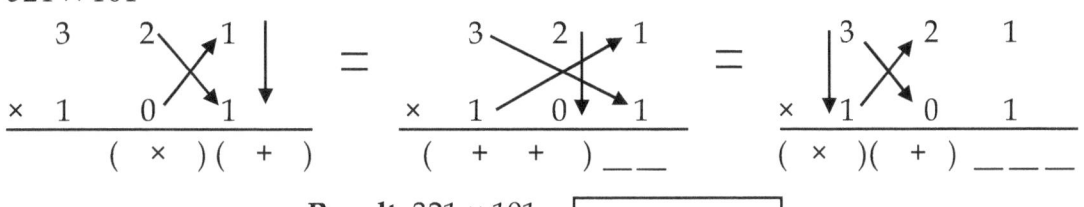

Result: $131 \times 220 = \boxed{}$

g. 321×101

Result: $321 \times 101 = \boxed{}$

h. 112×101

Result: $112 \times 101 = \boxed{}$

II. Multiply the below numbers using Urdhva Tiryagbhyam method

a. 321×121 b. 311×212 c. 220×311

d. 202×110 e. 202×220

EXERCISE 1.3

1. A large craton contains 102 small boxes. Each of the 102 boxes have 230 chocolates inside it. How many chocolates are there in the large box?

2. A library has 112 shelves. In each of these shelves there are 101 books. How many books are there in the library?

3. Ramesh has 204 boxes. In each of these boxes, he has 122 marbles. How many marbles does Ramesh have?

FAMOUS MATHEMATICIANS – BRAHMAGUPTA

Born:	c. 598 CE
Died:	after 665 CE
Nationality:	Ujjain, Indian

Famous for: He was an Indian Mathematician and Astronomer. First to give rules for computation with zero and treated zero as a number with its own right and not simply a decimal place holder. He gave solution for a general linear equation. Found the sum of the squares and cubes of first n integers. He marked his influence in geometry with his formula for cyclic quadrilaterals.

CHAPTER 2

- Urdhva Tiryagbhyam for 2-Digit Numbers with Carry Forward (Recap)

- Urdhva Tiryagbhyam for 3-Digit Numbers with Carry Forward

- Series of 9's

    - ❖ Multiplication of Numbers with Equal Number of 9's

- Learn Tables in Simple Way! – Nikhilam Method

URDHVA TIRYAGBHYAM

Meaning: Vertically and Crosswise

In the previous chapter we learnt the multiplication of any 2-digit and 3-digit numbers without carry. We proceed to learn the multiplication of 2-digit and 3-digit numbers with 'carry'.

URDHVA TIRYAGBHYAM FOR 2-DIGIT NUMBERS WITH CARRY FORWARD

We have seen how to multiply 2-digit numbers without carry. Let's discuss how to multiply 2-digit numbers with carry.

Let us consider an example, 13 × 24

Step 1: Multiply the digits in unit's place i.e., $3 \times 4 = 12$.

Carry is 1. We represent the carry with a star in the next digit, using Ekadhikena Purvena sutra.

$$
\begin{array}{ccc}
 & 1 & 3 \\
\times & \star 2 & 4 \\
\hline
 & & 2
\end{array}
$$

Step 2: Multiply the digits cross wise and add both products and add carry i.e.,

$(1 \times 4) + (3 \times 2) = 4 + 6 = 10 + 1 \text{ (carry)} = 11$

Carry is 1 and is represented as star.

$$
\begin{array}{ccc}
 & 1 & 3 \\
\times & \star 2 & 4 \\
\hline
 & 1 & 2
\end{array}
$$

Step 3: Multiply the digits in ten's place and add carry

$1 \times 2 = 2 + 1 \text{ (carry)} = 3$

$$
\begin{array}{ccc}
 & 1 & 3 \\
\times & 2 & 4 \\
\hline
3 & 1 & 2
\end{array}
$$

Result: 13 x 24 = 312.

All the 3 steps can be represented in a single step as follows,

Result: 3 1 2

Let us consider another example, 64 × 58

Step 1: Multiply the digits in unit's place

4 × 8 = 32. Here, carry is 3 (Using Ekadhikena Purvena Sutra). It is represented by star.

$$
\begin{array}{cc}
6 & 4 \\
\times \quad 5 & 8 \\
\hline
\star & \\
\hline
& 2 \\
\hline
\end{array}
$$

Step 2: Multiply the digits cross wise and add both products and add carry i.e.,

(6 × 8) + (4 × 5) = 48 + 20 = 68 + 3 (carry) = 71. Here 7 is carry and is represented by star.

$$
\begin{array}{cc}
6 & 4 \\
\times \quad 5 & 8 \\
\hline
\star & \\
\hline
1 & 2 \\
\hline
\end{array}
$$

Step 3: Multiply the ten's place digits and add carry i.e.,

6 × 5 = 30 + 7 (carry) = 37

$$
\begin{array}{ccc}
6 & & 4 \\
\times \quad 5 & & 8 \\
\hline
37 & 1 & 2 \\
\hline
\end{array}
$$

Result: 64 x 58 = 3712.

All the 3 steps can be represented in a single step as follows,

Result: 37 1 2

URDHVA TIRYAGBHYAM FOR 3-DIGIT NUMBERS WITH CARRY FORWARD

Here, we discuss how to multiply 3-digit numbers when there is carry.

Let us consider an example, 895 × 637

Step 1: Multiply the digits in unit's place

5 × 7 = 35

Carry will be 3 in the next step. This can be represented as:

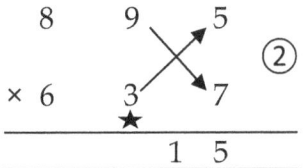

Step 2: Consider only the unit's and ten's place digits. Multiply the digits cross wise and add the products

$(9 \times 7) + (5 \times 3) = 63 + 15 = 78$. Carry 3 from the previous step to be added to 78.

$3 + 78 = 81$

Again, there is carry 8 to the next step

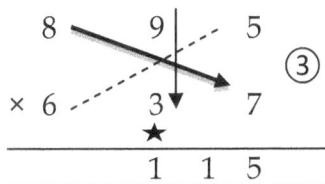

Step 3: Consider all the digits, Cross multiply and add the products.

$(8 \times 7) + (5 \times 6) + (9 \times 3) = 56 + 30 + 27 = 113$

Carry 8 from the previous step to be added to 113 i.e., $8 + 113 = 121$

Again, there is carry 12 to next step

$$\begin{array}{cccc} & 8 & 9 & 5 \\ \times & 6 & 3 & 7 \\ \hline & 1 & 1 & 5 \end{array} \quad ③$$

Step 4: Consider only the ten's and hundred's place digits. Multiply the digits cross wise and add the products

$(8 \times 3) + (9 \times 6) = 24 + 54 = 78$

Carry 12 from the previous step to be added to 78 i.e., $12 + 78 = 90$

Again, there is carry 9 to next step.

$$\begin{array}{ccccc} & 8 & 9 & 5 \\ \times & 6 & 3 & 7 \\ \hline 0 & 1 & 1 & 5 \end{array} \quad ④$$

Step 5: Multiply the digits in hundred's place

$(8 \times 6) = 48$

Carry 9 from the previous step to be added to 48 i.e., $9 + 48 = 57$

$$
\begin{array}{cccc}
 & 8 & 9 & 5 \\
\text{\textcircled{5}} \times & 6 & 3 & 7 \\
\hline
\textbf{Result: } 57 & 0 & 1 & 1 & 5
\end{array}
$$

Result: 895 x 637 = 570115.

Let us consider another example, 362 × 756

Step 1: Multiply the digits in unit's place i.e., $6 \times 2 = 12$

Carry will be 1 in the next step. This can be represented as,

$$
\begin{array}{cccc}
 & 3 & 6 & 2 \\
\times & 7 & 5 & 6 \;\text{\textcircled{1}} \\
\hline
 & & & 2
\end{array}
$$

Step 2: Consider only the unit's and ten's place digits. Multiply the digits cross wise and add the products

$(6 \times 6) + (2 \times 5) = 36 + 10 = 46$

Carry 1 from the previous step to be added to 46 i.e., $1 + 46 = 47$

Again, there is carry 4 to the next step

$$
\begin{array}{cccc}
 & 3 & 6 & 2 \\
\times & 7 & 5 & 6 \;\text{\textcircled{2}} \\
\hline
 & & 7 & 2
\end{array}
$$

Step 3: Consider all the digits. Cross multiply and add the products.

$(3 \times 6) + (2 \times 7) + (6 \times 5) = 18 + 14 + 30 = 62$

Carry 4 from the previous step to be added to 62 i.e., $4 + 62 = 66$

Again, there is carry 6 to next step

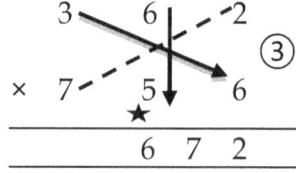

$$
\begin{array}{cccc}
 & 3 & 6 & 2 \\
\times & 7 & 5 & 6 \;\text{\textcircled{3}} \\
\hline
 & 6 & 7 & 2
\end{array}
$$

Step 4: Consider only the ten's and hundred's place digits. Multiply the digits crosswise and add the products

$(3 \times 5) + (6 \times 7) = 15 + 42 = 57$

Carry 6 from the previous step to be added to 57 i.e., $6 + 57 = 63$

Again, there is carry 6 to next step.

$$
\begin{array}{cccc}
3 & 6 & 2 \\
& ④ \\
\times \ 7 & 5 & 6 \\
\hline
& 3 & 6 & 7 & 2 \\
\end{array}
$$

Step 5: Multiply the digits in hundred's place

$(3 \times 7) = 21$

Carry 6 from the previous step to be added to 21

$6 + 21 = 27$

$$
\begin{array}{ccccc}
& 3 & 6 & 2 \\
⑤ \ \times 7 & 5 & 6 \\
\hline
27 & 3 & 6 & 7 & 2 \\
\end{array}
$$

Result: 362 × 756 = 273672.

SERIES OF 9'S

In this chapter, we learn multiplication with series of 9's i.e. the multiplier is in series of 9 (like 9, 99, 999, 9999) which is challenging, but we will learn an easy method.

MULTIPLICATION OF NUMBERS WITH EQUAL NUMBER OF 9'S

In this section, the multiplier and multiplicand have same number of digits and the multiplier is in series of 9.

Consider an example, 739 × 999.

$$
\begin{array}{rl}
739 & \rightarrow \text{Multiplicand} \\
\times\ 999 & \rightarrow \text{Multiplier} \\
\hline
\text{LHS} & |\ \text{RHS} \\
\end{array}
$$

Step 1: Find the LHS by subtracting 1 from the multiplicand. i.e. $739 - 1 = 738$

$$
\begin{array}{rl}
739 & \rightarrow \text{Multiplicand} \\
\times\ 999 & \rightarrow \text{Multiplier} \\
\hline
738 & |\ \text{RHS} \\
\end{array}
$$

Step 2: To find the RHS subtract each of the digit in the LHS from 9 individually and then write it in the place of RHS.

i.e. 9-7=2,

9-3=6,

And 9-8=1.

RHS will be 261.

$$739 \rightarrow \text{Multiplicand}$$
$$\times\ 999 \rightarrow \text{Multiplier}$$

738	261

Write the digits in LHS and RHS together we obtain the required answer.

Result: 739 × 999 = 738261.

Consider another example of 6835 x 9999

$$6835 \rightarrow \text{Multiplicand}$$
$$\times\ 9999 \rightarrow \text{Multiplier}$$

LHS	RHS

Step 1: Find the LHS by subtracting 1 from the multiplicand. i.e. 6835 -1 = 6834

$$6835 \rightarrow \text{Multiplicand}$$
$$\times\ 9999 \rightarrow \text{Multiplier}$$

6834	RHS

Step 2: To find the RHS subtract each of the digit in the LHS from 9 individually and then write it in the place of RHS.

i.e. 9 – 6 = 3

9 – 8 = 1

9 – 3 = 6

and 9 – 4 = 5.

RHS will be 3165.

$$6835 \rightarrow \text{Multiplicand}$$
$$\times\ 9999 \rightarrow \text{Multiplier}$$

6834	3165

Write the digits in LHS and RHS together we obtain the required answer.

Result: 6835 × 9999 = 68343165.

LEARN TABLES IN SIMPLE WAY!

MULTIPLICATION TABLES USING NIKHILAM METHOD

Now let us discuss how to remember tables using Nikhilam/Base Method. This method is very useful for those who find difficulty in remembering tables above 5. For using this method, one must be good in tables 2, 3, 4 and 5.

Now let us see how to form tables using Nikhilam Method.

Let us consider for example 8 x 6.

8	x	6
2		4

Here, we must consider PARAMA MITRA (complement) of 10. In this example, 8 and 6 are near to base number 10 with differences 2 and 4 respectively. It can be represented as shown in the table on left.

	8	x	6
	2		4
L.H.S =	4			= R.H.S

Now we must divide the answer into two parts, L.H.S and R.H.S.

The difference of numbers cross-wise should be found now. It can be 8 – 4 or 6 – 2 as both gives same answer 4. Let us write the result in the L.H.S digit.

	8	x	6
	2		4
L.H.S =	4		8	= R.H.S

Now, to get the R.H.S digit, we must multiply the differences of numbers i.e. 2 x 4 = 8. This goes to the R.H.S part of the answer

Result: 8 x 6 = 48.

EXERCISE 2.1

I. Multiply the below numbers using Urdhva Tiryagbhyam method

a. 29×33

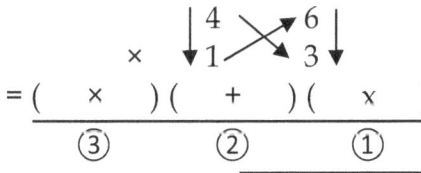

= (×   ) (+   ) (×   )
  ③   ②   ①

Result: $29 \times 33 =$ ☐

b. 15×52

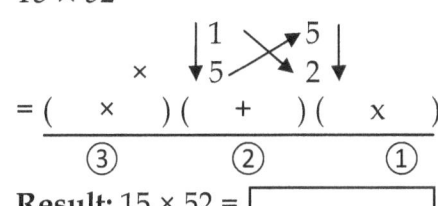

= (×   ) (+   ) (×   )
  ③   ②   ①

Result: $15 \times 52 =$ ☐

c. 46×13

= (×   ) (+   ) (×   )
  ③   ②   ①

Result: $46 \times 13 =$ ☐

d. 45×78

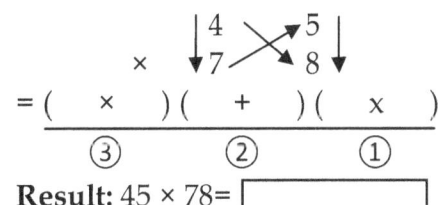

= (×   ) (+   ) (×   )
  ③   ②   ①

Result: $45 \times 78 =$ ☐

e. 56×34

= (×   ) (+   ) (×   )

Result: $56 \times 34 =$ ☐

f. 31×14

= (×   ) (+   ) (×   )

Result: $31 \times 14 =$ ☐

g. 24×32

= (×)(+)(x)

Result: 24 × 32 = ☐

h. 59×12

= (×)(+)(x)

Result: 59 × 12= ☐

II. **Multiply the below numbers using Urdhva Tiryagbhyam method**

a. 16×33 b. 61×36 c. 31×29

d. 76×54 e. 19×37

EXERCISE 2.2

I. **Multiply the below numbers using Urdhva Tiryagbhyam method**

a. 564×311

Result: 564 × 311 = ☐

b. 381×521

Result: 381 × 521 = ☐

c. 895×756

Result: 895 × 756 = ☐

d. 362×637

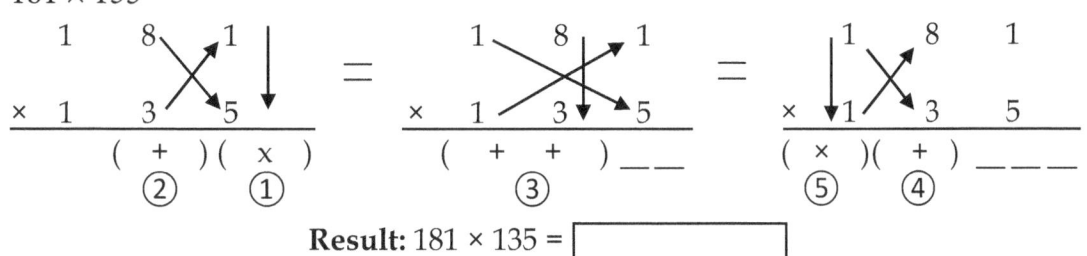

Result: $362 \times 637 =$ ☐

e. 181×135

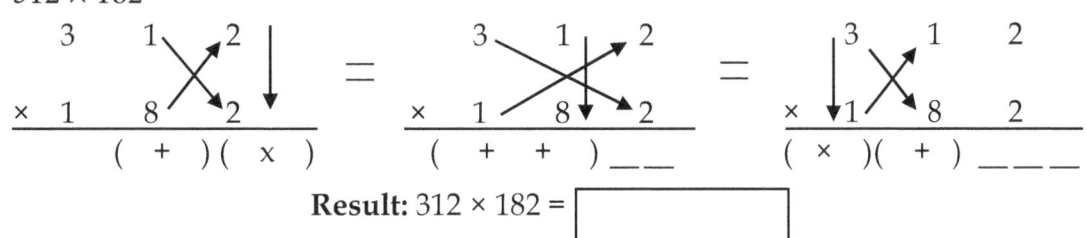

Result: $181 \times 135 =$ ☐

f. 312×182

Result: $312 \times 182 =$ ☐

g. 261×136

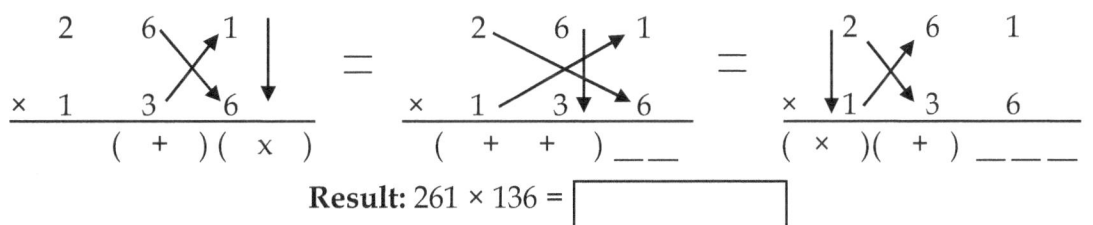

Result: $261 \times 136 =$ ☐

h. 911×131

Result: $911 \times 131 =$ ☐

II. Multiply the below numbers using Urdhva Tiryagbhyam method

a. 339×532 b. 675×456 c. 366×135

d. 915×531

EXERCISE 2.3

I.   Solve the following (With equal number of 9's):

a. 34×99

  LHS = _____

  RHS = _____

  Result: $34 \times 99 =$ ☐

b. 765×999

  LHS = _____

  RHS = _____

  Result: $765 \times 999 =$ ☐

c. 347×999

  LHS = _____

  RHS = _____

  Result: $347 \times 999 =$ ☐

d. 124×999

  LHS = _____

  RHS = _____

  Result: $124 \times 999 =$ ☐

e. 8634×9999

  LHS = _____

  RHS = _____

  Result: $8634 \times 9999 =$ ☐

f. 5435×9999

  LHS = _____

  RHS = _____

  Result: $5435 \times 9999 =$ ☐

II.   Solve the following:

a. 145×999

b. 76532432×99999999

c. 3879×9999

d. 12×99

e. 7532×9999

f. 134×999

EXERCISE 2.4

a.   There are 177 toy shops on one street. Each toy shop has 139 toys in it. How many toys are there in total?

b.   One large box has 124 small boxes. There are 351 such large boxes. How many small boxes are there?

c.   A room requires 186 bricks to be built. 123 rooms should be built. How many bricks are required?

FAMOUS MATHEMATICIANS – BLAISE PASCAL

Born:	19 June 1623
Died:	19 August 1662
Nationality:	French

Famous for: He was a French mathematician, physicist, philosopher. He made major contributions in the study of fluids, further clarifying details about pressure, vacuum. Was one of the first of two inventors of mechanical calculators and developed probability theory. He developed Pascal's triangle which is representation for binomial coefficients. He also invented the syringe and hydraulic press.

CHAPTER 3

- Nikhilam Sutra for Multiplication (Recap)

  - ❖ When Numbers Are below the Base

  - ❖ When Numbers Are above the Base

  - ❖ When Number of Digits in Rhs Exceeds Number of Zeros in Base

  - ❖ When One Number Is above the Base and Another Number Is below the Base

- Multiplication Using Nikhilam Sutra for Numbers with Different Base

- Series of 9's

  - ❖ Multiplication of Numbers with Higher Number of 9's

- Learn Tables in Simple Way! – Trick for Table 8

NIKHILAM SUTRA FOR MULTIPLICATION

Nikhilam Sutra/Base method is an easy technique in Vedic Mathematics for multiplication of numbers.

Multiplication using Nikhilam Sutra is done when numbers are closer to power of 10 or base numbers.

What are base numbers?

Powers of 10 such as 10, 100, 1000, 10000 etc. are called base numbers.

In this chapter, we shall recall the topics discussed in Level-1 about Nikhilam Sutra and other relevant concepts. Listed below are the topics related to the multiplication methods using Nikhilam Sutra:

- When numbers are below the base
- When numbers are above the base
- When number of digits in RHS exceeds number of zeros in base.
- When one number is above the base and another number is below the base

Let us discuss all the concepts in detail. To understand the base method, we need to observe first whether the numbers are near to the base numbers.

Let us consider few examples to learn whether the numbers are close to base numbers.

1. Multiply 6 x 8 – we can use base method, since both numbers are closer to base number 10. This can be represented as

$$6 \ (10) \longrightarrow -4$$
$$8 \ (10) \longrightarrow -2$$
_____

2. Multiply 103 x 105 – we can use base method, since both numbers are closer to base number 100. This can be represented as

$$103 \ (100) \longrightarrow +3$$
$$105 \ (100) \longrightarrow +5$$
_____

Now, we shall discuss the Nikhilam method in detail.

MULTIPLICATION USING BASE METHOD

1. WHEN BOTH NUMBERS ARE BELOW BASE

This method can be used when both the numbers are lesser than the nearest Base number.

Let us consider an example, 93 x 94

Step 1: Here the base is 100, since the numbers 93 and 94 are closer to 100. This can be represented as: 93 (100)

94 (100)

Step 2: We can observe here that both the numbers are below base. Now we must find the difference between the numbers and mention the differences. i.e. 100 – 93 = 7 and 100 – 94 = 6.

So, the differences are 7 and 6. This can be represented as:

93 (100) ⟶ - 7
94 (100) ⟶ - 6
Answer: _____

NOTE: We write '–' before the difference to indicate that the number is less than the base number.

Step 3: Divide the answer into LHS and RHS

93 (100) ⟶ - 7
94 (100) ⟶ - 6
Answer: LHS | RHS

NOTE: Here LHS and RHS are separated by straight line.

Step 4: Note that the number of digits in RHS = Number of zeros in base: In this case, the base is 100, the number of zeros in base is 2, hence the number of digits in RHS must be 2. Here RHS is represented by 2 blanks.

93 (100) ⟶ - 7
94 (100) ⟶ - 6
Answer: LHS | _ _

Step 5: Now multiply the differences and write it in RHS.

In this case the difference is 7 and 6. So 7 x 6 = 42. So RHS will be 42

```
                    93 (100) ——➤ - 7
                    94 (100) ——➤ - 6
      Answer:    LHS  | 4 2
```

NOTE: We have RHS of the answer.

Step 6: To find LHS we perform cross subtraction (since both the numbers are below the base) and write it in LHS. Please note that the answer is same for both cross subtractions.

i.e. 93 – 6 = 87 and 94 – 7 = 87. Here in both case answer is 87. Hence LHS is 87.

```
                    93 (100) ——➤ - 7
                    94 (100) ——➤ - 6
      Answer:    8 7  | 4 2
```

Result: 93 x 94 = 8742.

Let us consider another example 994 x 998

Step 1: Here the base is 1000, since the numbers 994 and 998 are closer to 1000 and here both the numbers are below base.

994 (1000)

998 (1000)

Step 2: Now we must find the difference between the numbers and mention the differences.

Base = 1000. Difference = base – number i.e. 1000 – 994 = 6 and 1000 – 998 = 2

So, the differences are 6 and 2. This can be represented as:

```
                    994 (1000) ——➤ - 6
                    998 (1000) ——➤ - 2
      Answer:    _____
```

Step 3: Divide the answer into LHS and RHS

```
                    994 (1000) ——➤ - 6
                    998 (1000) ——➤ - 2
      Answer:    LHS  |  RHS
```

NOTE: Here LHS and RHS are separated by straight line

Step 4: Note that the number of digits in RHS = Number of zeros in base: In this case, the base is 1000, the number of zeros in base is 3, hence the number of digits in RHS must be 3. Here RHS is represented by 3 blanks.

```
                    994 (1000) ——➤ - 6
                    998 (1000) ——➤ - 2
      Answer:    LHS  |  _ _ _
```

Step 5: Now multiply the differences and write it in RHS.

In this case the difference is 6 and 2. So 6 x 2 = 12. So RHS will be 12

$$
\begin{array}{r}
994\ (1000) \longrightarrow -6 \\
998\ (1000) \longrightarrow -2 \\
\hline
\text{Answer:} \quad \text{LHS} \mid 0\ 1\ 2 \\
\end{array}
$$

NOTE: We have RHS of the answer.

Step 6: To find LHS we perform cross subtraction (since both the numbers are below the base) and write it in LHS. Please note that the answer is same for both cross subtractions.

i.e. 994 – 2 = 992 and 998 – 6 = 992. Here in both case answer is 992. Hence LHS is 992.

$$
\begin{array}{r}
994\ (1000) \longrightarrow -6 \\
998\ (1000) \longrightarrow -2 \\
\hline
\text{Answer:} \quad 9\ 9\ 2 \mid 0\ 1\ 2 \\
\end{array}
$$

Result: 994 x 998 = 992012.

2. WHEN BOTH THE NUMBERS ARE ABOVE THE BASE

This method of multiplication is used when the numbers are greater than the nearest base number.

Let us consider an example 101 x 104.

Step 1: Here the base is 100, since the numbers 101 and 104 are closer to 100 and here both the numbers are above the base.

101 (100)

104 (100)

Step 2: Now we must find the difference between the numbers and mention the differences.

So, the differences are 1 and 4. This can be represented as:

$$
\begin{array}{r}
101\ (100) \longrightarrow +1 \\
104\ (100) \longrightarrow +4 \\
\hline
\end{array}
$$

NOTE: We write '+' before the difference to indicate that the number is greater than the base number.

Step 3: Divide the answer into LHS and RHS

$$
\begin{array}{r}
101\ (100) \longrightarrow +1 \\
104\ (100) \longrightarrow +4 \\
\hline
\text{Answer:} \quad \text{LHS} \mid \text{RHS} \\
\end{array}
$$

> *NOTE:* Here LHS and RHS are separated by straight line.

Step 4: Note that the number of digits in RHS = Number of zeros in base: In this case, the base is 100, the number of zeros in base is 2, hence the number of digits in RHS must be 2. Here RHS is represented by 2 blanks.

$$101 \ (100) \longrightarrow +1$$
$$104 \ (100) \longrightarrow +4$$

Answer: ___LHS___|___ _ _ ___

Step 5: Now multiply the differences and write it in RHS.

In this case the differences are 1 and 4. So 1 x 4 = 4. So RHS will be 04

$$101 \ (100) \longrightarrow +1$$
$$104 \ (100) \longrightarrow +4$$

Answer: ___LHS___|___0 4___

> *NOTE:* We have RHS of the answer.

Step 6: To find LHS we perform cross addition (since both the numbers are above the base). Please note that the answer is same for both cross additions.

i.e. In this case: 101+ 4 = 105 and 104 +1 = 105. So, the LHS = 105

$$101 \ (100) \longrightarrow +1$$
$$104 \ (100) \longrightarrow +4$$

Answer: ___1 0 5___|___0 4___

Result: 101 x 104 = 10504.

3. WHEN NUMBER OF DIGITS IN RHS EXCEEDS NUMBER OF ZEROS IN BASE

This method of multiplication is used when number of digits in the multiplication of the differences (RHS) is greater than the number of zeros in base.

It can be represented as number of digits of the answer > number of zeros in the base In such case, we use the carry forward concept.

Let us consider an example 930 x 960.

Step 1: Here the base is 1000, since the numbers 930 and 960 are closer to 1000 and here both the numbers are below base.

$$930 \ (1000)$$

$$960 \ (1000)$$

Step 2: Now we must find the difference between the numbers and mention the differences. So, the differences are 70 and 40. This can be represented as:

$$930(1000) \longrightarrow -70$$
$$960(1000) \longrightarrow -40$$
Answer: _____

Step 3: Divide the answer into LHS and RHS

$$930(1000) \longrightarrow -70$$
$$960(1000) \longrightarrow -40$$

Answer:	LHS	RHS

NOTE: Here LHS and RHS are separated by straight line.

Step 4: Note that the number of digits in RHS = Number of zeros in base: In this case, as base is 1000, the number of zeros in base is 3, hence the number of digits in RHS must be 3. Here RHS is represented by 3 blanks.

$$930(1000) \longrightarrow -70$$
$$960(1000) \longrightarrow -40$$

Answer:	LHS	_ _ _

NOTE: Three empty blank in RHS represent 3 digits of the answer.

Step 5: Now multiply the differences and write it in RHS.

In this case differences are 70 and 40. So, 70 x 40 = 2800, which is 4-digit answer.

Please note the number of digits in RHS is 3, whereas we have 4-digit answer. In such case carry over 2 and write 800 in RHS. This can be represented as:

$$930(1000) \longrightarrow -70$$
$$960(1000) \longrightarrow -40$$
$$\text{Carry over} \quad \rightarrow \quad 2$$

Answer:	LHS	800

Step 6: To find LHS we perform cross subtraction and add the carry over and write it in LHS.

i.e. 930 – 40 = 890 and 960 – 70 = 890.

So LHS = 890 + 2 (carry over) = 892.

LHS is 892. This can be represented as:

$$930(1000) \longrightarrow -70$$
$$960(1000) \longrightarrow -40$$

Answer:	892	800

Result: 930 × 960 = 892800.

Let us consider another example, 1080 x 1020.

Here, when we observe the numbers, both are above the base number.

Step 1: Find the base. Here the nearest base is 1000, and both the numbers are above base. This can be represented as:

$$1080(1000)$$

$$1020(1000)$$

Step 2: Now we must find the difference between the numbers and mention the differences.

So, the differences are 80 and 20. This can be represented as:

$$1080(1000) \longrightarrow +80$$
$$1020(1000) \longrightarrow +20$$

Answer: _____

Step 3: Divide the answer in LHS and RHS

$$1080(1000) \longrightarrow +80$$
$$1020(1000) \longrightarrow +20$$

Answer: | LHS | RHS |

NOTE: Here LHS and RHS are separated by straight line.

Step 4: The number of digits in RHS = Number of zeros in base

In this case number of zeros in base = 3 so the number of digits in RHS must be 3. This can be represented as:

$$1080(1000) \longrightarrow +80$$
$$1020(1000) \longrightarrow +20$$

Answer: | LHS | _ _ _ |

Step 5: Now multiply the differences and write it in RHS.

In this case differences are 80 & 20. So, 80 x 20 = 1600, which is 4-digit answer.

Please note the number of digits in RHS is 3, whereas we have 4-digit answers. So, in such case carry over 1 and write 600 in RHS. This can be represented as:

$$1080(1000) \longrightarrow +80$$
$$1020(1000) \longrightarrow +20$$
$$\text{Carry over} \quad \rightarrow \quad 1$$

Answer: | LHS | 600 |

Step 6: To find LHS we perform cross addition (since the numbers are above base) and add the carry over and write it in LHS.

i.e. 1080 + 20 = 1100 and 1020 + 80 = 1100.

So LHS = 1100 + 1 (carry over) = 1101.

LHS is 1101. This can be represented as:

$$1080(1000) \longrightarrow +80$$
$$1020(1000) \longrightarrow +20$$

Answer:	1101	600

Result: 1080 × 1020 = 1101600.

4. WHEN ONE NUMBER IS ABOVE BASE AND ANOTHER NUMBER IS BELOW BASE

This method of multiplication is used when one of the number is above the base and other is below the base.

Let us consider an example 107 x 98.

Step 1: Here the base is 100. 107 is above the base and 98 is below the base.

$$107\,(100)$$

$$98\,(100)$$

Step 2: Now we must find the difference between the numbers and mention the differences.

So, the differences are 7 and 2. This can be represented as:

$$107(100) \longrightarrow +7$$
$$98(100) \longrightarrow -2$$

Answer:

Step 3: Divide the answer in LHS and RHS

$$107(100) \longrightarrow +7$$
$$98(100) \longrightarrow -2$$

Answer:	LHS	RHS

NOTE: Here LHS and RHS are separated by straight line.

Step 4: Note that the number of digits in RHS = Number of zeros in base: In this case, as base is 100, the number of zeros in base is 2, hence the number of digits in RHS must be 2. Here RHS is represented by 2 blanks.

$$107(100) \longrightarrow +7$$
$$98(100) \longrightarrow -2$$

Answer:	LHS	_ _

Step 5: Now multiply the differences and write it in RHS.

In this case the differences are 7 and 2. So +7 x -2 = -14. So RHS will be -14

$$107(100) \longrightarrow +7$$
$$98(100) \longrightarrow -2$$

Answer:	LHS	- 14

NOTE: We have RHS of the answer.

Step 6: To find LHS we perform cross calculation and write it in LHS. Please note that the answer is same for both cross operations.

i.e. In this case: 107 - 2 = 105 and 98 +7 = 105. So, the LHS = 105

$$107(100) \longrightarrow +7$$
$$98(100) \longrightarrow -2$$
$$\text{Answer:} \quad \overline{105 \mid -14}$$

Step 7: Multiply LHS with the base i.e. 105 x 100 = 10500

Now add RHS to 10500 i.e. 10500 +(–14) = 10500-14 = 10486.

Result: 107 x 98 = 10486.

NOTE: Step 7 can be written as (LHS x base) + RHS = Final answer.

Let us consider another example 1006 x 998.

Step 1: Here the base is 1000. 1006 is above the base and 998 is below the base.

$$1006 \ (1000)$$

$$998 \ (1000)$$

Step 2: Now we must find the difference between the numbers and mention the differences.

So, the differences are 6 and 2. This can be represented as:

$$1006(1000) \longrightarrow +6$$
$$998(1000) \longrightarrow -2$$
$$\text{Answer:} \quad \overline{}$$

Step 3: Divide the answer in LHS and RHS

$$1006(1000) \longrightarrow +6$$
$$998(1000) \longrightarrow -2$$
$$\text{Answer:} \quad \overline{\text{LHS} \mid \text{RHS}}$$

NOTE: Here LHS and RHS are separated by straight line.

Step 4: Note that the number of digits in RHS = Number of zeros in base: In this case, as base is 1000, the number of zeros in base is 3, hence the number of digits in RHS must be 3. Here RHS is represented by 3 blanks.

$$1006(1000) \longrightarrow +6$$
$$998(1000) \longrightarrow -2$$
$$\text{Answer:} \quad \overline{\text{LHS} \mid \_\,\_\,\_}$$

Step 5: Now multiply the differences and write it in RHS.

In this case the differences are 6 and 2. So +6 x -2 = -12. So RHS will be -12

$$1006(1000) \longrightarrow +6$$
$$998(1000) \longrightarrow -2$$

Answer: | LHS | -012 |

Step 6: To find LHS we perform cross calculation and write it in LHS. Please note that the answer is same for both cross operations.

i.e. In this case: 1006 – 2 = 1004 and 998 + 6 = 1004. So, the LHS = 1004

$$1006(1000) \longrightarrow +6$$
$$998(1000) \longrightarrow -2$$

Answer: | 1004 | -012 |

Step 7: Multiply LHS with the base i.e. 1004 x 1000 = 1004000

Now add RHS to 1004000 i.e. 1004000 +(–0 1 2) = 1004000-012 = 1003988.

NOTE: Step 7 can be written as (LHS x base) + RHS = Final answer.

Result: 1006 x 998 =1003988.

MULIPLICATION OF NUMBERS WITH DIFFERENT BASE

In the first section we have seen, how to multiply the numbers of same base using base method. Now we will see the multiplication of numbers with different base. Let us consider one example to explain this in detail.

Multiply 79 x 950.

Step 1: 79 is the lowest number and to make same base as 950 multiply 79 with 10.

i.e. 79 x 10 = 790.

Now base of 790 is 1000. Hence base of both the numbers are same i.e. 1000

$$790 \ (1000)$$

$$950 \ (1000)$$

Step 2: Now we must find the difference between the numbers and mention the differences.

So, the differences are 210 and 50. This can be represented as:

$$790(1000) \longrightarrow -210$$
$$950(1000) \longrightarrow -\ 50$$

Answer:

Step 3: Divide the answer as LHS and RHS.

$$790(1000) \longrightarrow -210$$
$$950(1000) \longrightarrow -\ 50$$

Answer: | LHS | RHS |

NOTE: Here LHS and RHS are separated by straight line.

Step 4: Note that the number of digits in RHS = Number of zeros in base: In this case, as base is 1000, the number of zeros in base is 3, hence the number of digits in RHS must be 3. Here RHS is represented by 3 blanks.

$$790(1000) \longrightarrow -210$$
$$950(1000) \longrightarrow -\ 50$$
Answer: $\underline{\quad\ \text{LHS}\quad\ |\ \_\ \_\ \_\quad}$

Step 5: Now multiply the differences and write it in RHS.

In this case the differences are 210 and 50. So 210 x 50 = 10500. which is 5-digit answer. Please note the number of digits in RHS is 3, whereas we have 5-digit answers. So, in such case carry over 10 and write 500 in RHS. This can be represented as:

$$790(1000) \longrightarrow -210$$
$$950(1000) \longrightarrow -\ 50$$
$$\text{Carry over} \quad \rightarrow \quad 10$$
Answer: $\underline{\quad\ \text{LHS}\quad\ |\ 500\quad}$

Step 6: To find LHS we perform cross subtraction (since the numbers are below base) and add the carry over and write it in LHS.

i.e. 790 – 50 = 740 or 950 – 210 = 740.

So LHS = 740 +10 (carry over) = 750.

So LHS is 750 and now write it in LHS. This can be represented as:

$$790(1000) \longrightarrow -210$$
$$950(1000) \longrightarrow -\ 50$$
Answer: $\underline{\quad 750\quad |\ 500\quad}$

We have answer as, 750500.

To get the result, divide the answer by same power of 10 i.e. in this case we have multiplied 79 by 10, so divide the answer 750500 by 10 i.e. 750500/10=75050

Result: 79 x 950 = 75050.

Let us consider another example 22 x 350.

Step 1: 22 is the lowest number and to make same base as 350 multiply 22 with 10.

i.e. 22 x 10 = 220.

Now base of 220 is 100. Hence base of both the numbers are same i.e. 100

$$220\ (100)$$

$$350\ (100)$$

Step 2: Now we must find the difference between the numbers and mention the differences.

So, the differences are 120 and 250. This can be represented as:

$$220(100) \longrightarrow + 120$$
$$350(100) \longrightarrow + 250$$

Answer: _____

Step 3: Divide the answer as LHS and RHS.

$$220(100) \longrightarrow + 120$$
$$350(100) \longrightarrow + 250$$

Answer: | LHS | RHS |

NOTE: Here LHS and RHS are separated by straight line.

Step 4: Note that the number of digits in RHS = Number of zeros in base: In this case, as base is 100, the number of zeros in base is 2, hence the number of digits in RHS must be 2. Here RHS is represented by 2 blanks.

$$220(100) \longrightarrow + 120$$
$$350(100) \longrightarrow + 250$$

Answer: | LHS | _ _ |

Step 5: Now multiply the differences and write it in RHS.

In this case the differences are 120 and 250. So, 120 x 250 = 30000. which is 5-digit answer. Please note the number of digits in RHS is 2, whereas we have 5-digit answers. So, in such case carry over 300 and write 00 in RHS. This can be represented as:

$$220(100) \longrightarrow + 120$$
$$350(100) \longrightarrow + 250$$
Carry over $\rightarrow$ 300

Answer: | LHS | 00 |

Step 6: To find LHS we perform cross addition (since the numbers are above base) and add the carry over and write it in LHS.

i.e. 220 + 250 = 470 or 350 + 120 = 470.

So LHS = 470 +300 (carry over) = 770.

So LHS is 770. This can be represented as:

$$220(100) \longrightarrow + 120$$
$$350(100) \longrightarrow + 250$$

Answer: | 770 | 00 |

We have answer as, 77000.

To get the final answer, divide the answer by same power of 10 i.e. in this case we have multiplied 22 by 10,so divide the answer 77000 by 10 i.e. 77000/10=7700

Result: 22 x 350 = 7700.

MULTIPLICATION OF NUMBERS WITH HIGHER NUMBER OF 9'S

The previous method discussed in Chapter 2 can be followed only when the digits in the multiplier and multiplicand are equal. If the digits in the multiplier are greater than the digits in the multiplicand we must follow another method which we will be discussed in this section.

Consider an example, 254 × 9999

Here the multiplier has 4 digits whereas the multiplicand has only 3 digits.

Step 1: Make the number of digits in the multiplicand equal to the digits in the multiplier by placing 0 (zero) before the multiplicand.

Here, the multiplier has 4 digits and the multiplicand has 3 digits so just add one zero before 254 i.e., 0254. Now the digits are equal to the number of 9's in the multiplier and follow the previous method to find the answer.

$$0254 \longrightarrow \text{Multiplicand}$$
$$\times\ 9999 \longrightarrow \text{Multiplier}$$
$$\overline{\qquad \text{LHS} \mid \text{RHS} \qquad}$$

Step 2: Find the LHS by subtracting 1 from the multiplicand. i.e. 254 -1 = 253

$$0254 \longrightarrow \text{Multiplicand}$$
$$\times\ 9999 \longrightarrow \text{Multiplier}$$
$$\overline{\qquad 253 \mid \text{RHS} \qquad}$$

Step 3: To find the RHS subtract each of the digit in the LHS from 9 individually and then write it in the place of RHS

i.e.   9 – 2 = 7,

9 – 5 = 4,

And 9 – 3 = 6.

RHS will be 746.

$$0254 \longrightarrow \text{Multiplicand}$$
$$\times\ 9999 \longrightarrow \text{Multiplier}$$
$$\overline{\qquad 253 \mid 746 \qquad}$$

Step 4: Since we have placed a zero before 254 to make it a 4-digit number in Step 1 now we must modify the answer accordingly. So, we place a 9 between the digits of LHS and RHS.

$$0254 \longrightarrow \text{Multiplicand}$$
$$\times\ 9999 \longrightarrow \text{Multiplier}$$
$$\overline{\qquad 253 \boxed{9} 746 \qquad}$$

Write the digits in LHS and RHS together we obtain the required answer.

Result: 254 × 9999 = 2539746.

Consider an example 12345 x 9999999

Here the multiplier has 7 digits whereas the multiplicand has only 5 digits.

Step 1: Make the number of digit in the multiplicand equal to the digits in the multiplier by placing 0 (zero) before the multiplicand.

Here, the multiplier has 7 digits and the multiplicand has 5 digits so just add two zeros before 12345 i.e., 0012345. Now the digits are equal to the number of 9's in the multiplier and follow the previous method to find the answer.

$$0012345 \longrightarrow \text{Multiplicand}$$
$$\times\ 9999999 \longrightarrow \text{Multiplier}$$

LHS	RHS

Step 2: Find the LHS by subtracting 1 from the multiplicand. i.e. 12345 -1 = 12344

$$0012345 \longrightarrow \text{Multiplicand}$$
$$\times\ 9999999 \longrightarrow \text{Multiplier}$$

12344	RHS

Step 3: To find the RHS subtract each of the digit in the LHS from 9 individually and then write it in the place of RHS

i.e. 9 – 1 = 8,

9 – 2 = 7,

9 – 3 = 6,

9 – 4 = 5,

and 9 – 4 = 5.

RHS will be 87655.

$$0012345 \longrightarrow \text{Multiplicand}$$
$$\times\ 9999999 \longrightarrow \text{Multiplier}$$

12344	87655

Step 4: Since we have placed two zeros before 12345 to make it a 7-digit number in Step 1 now we must modify the answer accordingly. So, we place a 99 between the digits of LHS and RHS.

$$0012345 \longrightarrow \text{Multiplicand}$$
$$\times\ 9999999 \longrightarrow \text{Multiplier}$$

12344	**99**	87655

Write the digits in LHS and RHS together we obtain the required answer.

Result: 12345 × 9999999 = 123449987655.

> **NOTE:** If the multiplier has more digits than the multiplicand we place zeros before the multiplicand to get equal number of digits in the multiplier and multiplicand. The number of zeros inserted in 1st step is equal to the number of 9's inserted in the last step.

LEARN TABLES IN SIMPLE WAY!

TRICK OF MULTIPLICATION TABLE FOR 8

Here is an easy and quick trick to remember multiplication table of 8.

0	8
1	6
2	4
3	2
4	0
4	8
5	6
6	4
7	2
8	0

To get this table, just make 2 columns and write numbers from 0 to 8 in one column, repeating 4 twice as shown below. In second column, write the even numbers below 10 in reverse order. i.e. numbers 8,6,4,2,0 and repeat again as shown on the left.

8 x 1	=	08
8 x 2	=	16
8 x 3	=	24
8 x 4	=	32
8 x 5	=	40
8 x 6	=	48
8 x 7	=	56
8 x 8	=	64
8 x 9	=	72
8 x 10	=	80

Have a look at the figure on right.

Now we have written the multiplication table of 8.

Let us join both the columns together and the multiplication table of 8 is formed! Have a look at the complete table on right.

Let us write the multiplication table of 8 table once again.

	COLUMN 1	COLUMN 2
8 x 1 =		
8 x 2 =		
8 x 3 =		
8 x 4 =		
8 x 5 =		
8 x 6 =		
8 x 7 =		
8 x 8 =		
8 x 9 =		
8 x 10 =		

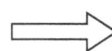

8 x 1	=	
8 x 2	=	
8 x 3	=	
8 x 4	=	
8 x 5	=	
8 x 6	=	
8 x 7	=	
8 x 8	=	
8 x 9	=	
8 x 10	=	

EXERCISE 3.1

I. **Find the nearest base number, whether it is below or above the base and represent it for the following**

a. 86

The nearest base of 86 is ☐

Is it **below/above** the base? ☐

It can be represented as

86 (_____) → ☐

b. 108

The nearest base of 108 is ☐

Is it **below/above** the base? ☐

It can be represented as

108 (_____) → ☐

c. 453

The nearest base of 453 is ☐

Is it **below/above** the base? ☐

It can be represented as

453 (_____) → ☐

d. 1009

The nearest base of 1009 is ☐

Is it **below/above** the base? ☐

It can be represented as

1009 (_____) → ☐

e. 9999

The nearest base of 9999 is ☐

Is it **below/above** the base? ☐

It can be represented as

9999 (_____) → ☐

f. 908

The nearest base of 908 is ☐

Is it **below/above** the base? ☐

It can be represented as

908(_____) → ☐

g. 45

The nearest base of 45 is ☐

Is it **below/above** the base? ☐

It can be represented as

45 (_____) → ☐

h. 02

The nearest base of 02 is ☐

Is it **below/above** the base? ☐

It can be represented as

02 (_____) → ☐

II.   Solve by base method:

a.   98 x 94

Numbers are below/above base

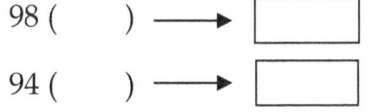

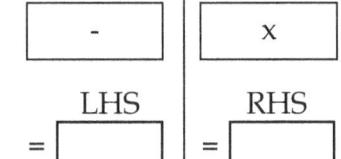

| - | x |

LHS | RHS

= [   ] | = [   ]

Result: 98 x 94 = [   ]

b.   106 x 104

Numbers are below/above base

106 (   ) ⟶ [   ]

104 (   ) ⟶ [   ]

| - | x |

LHS | RHS

= [   ] | = [   ]

Result: 106 x 104 = [   ]

c.   12 x 14

Numbers are below/above base

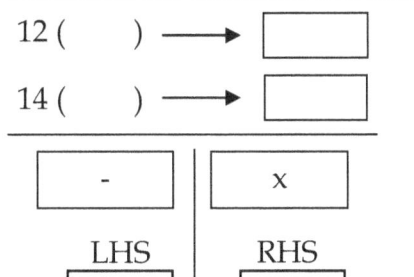

12 (   ) ⟶ [   ]

14 (   ) ⟶ [   ]

| - | x |

LHS | RHS

= [   ] | = [   ]

Result: 12 x 14= [   ]

d.   988 x 996

Numbers are below/above base

988 (   ) ⟶ [   ]

996 (   ) ⟶ [   ]

| - | x |

LHS | RHS

= [   ] | = [   ]

Result: 988 x 996 = [   ]

e.   102 x 108

Numbers are below/above base

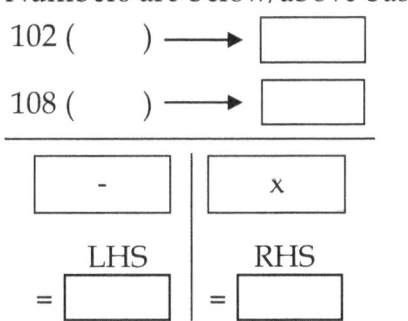

102 (   ) ⟶ [   ]

108 (   ) ⟶ [   ]

| - | x |

LHS | RHS

= [   ] | = [   ]

Result: 102 x 108 = [   ]

f.   992 x 999

Numbers are below/above base

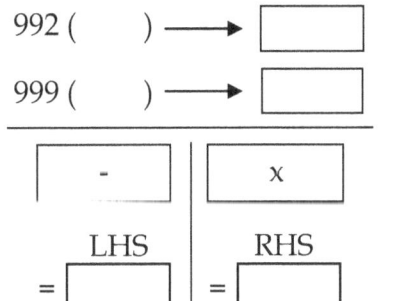

992 (   ) ⟶ [   ]

999 (   ) ⟶ [   ]

| - | x |

LHS | RHS

= [   ] | = [   ]

Result: 992 x 999 = [   ]

III.   Solve by base method:

a.   108 x106

b.   99 x 91

c.   105 x 101

d.   994 x 899

e.   998 x 990

f.   99 x 97

g.   997 x 990

EXERCISE 3.2

I. **Solve by base method: (Digits in RHS exceeds the number of zeros in base)**

a. 890 x 880

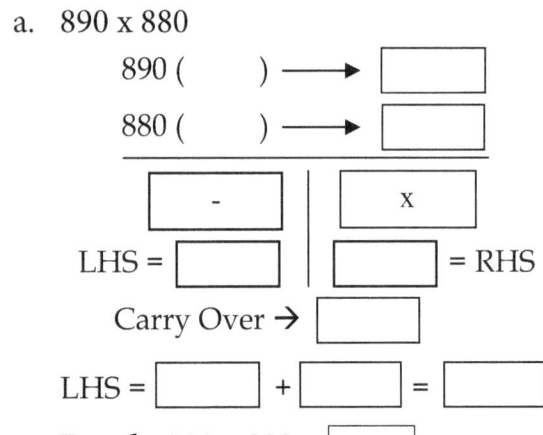

    Result: 890 x 880 =

b. 78 x 90

890 ( ) ⟶

90 ( ) ⟶

LHS = ‖ = RHS

Carry Over →

LHS = + =

Result: 78 x 90 =

c. 289 x 386

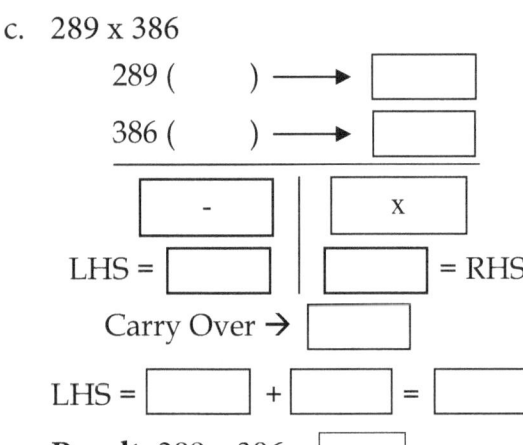

    Result: 289 x 386 =

d. 800 x 887

800 ( ) ⟶

887 ( ) ⟶

LHS = ‖ = RHS

Carry Over →

LHS = + =

Result: 800 x 887 =

e. 1021 x 1018

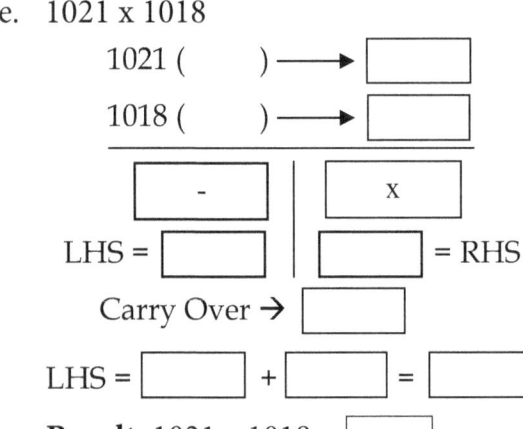

    Result: 1021 x 1018 =

f. 678 x 980

678 ( ) ⟶

980 ( ) ⟶

LHS = ‖ = RHS

Carry Over →

LHS = + =

Result: 678 x 980 =

II. Solve by base method:

a. 650 x 880

b. 168 x 180

c. 682 x 790

d. 786 x 991

e. 220 x 350

f. 900 x 960

g. 120 x 150

EXERCISE 3.3

I. Solve by base method: (One number is above base and another number is below base)

a. 102 x 99

b. 1006 x 997

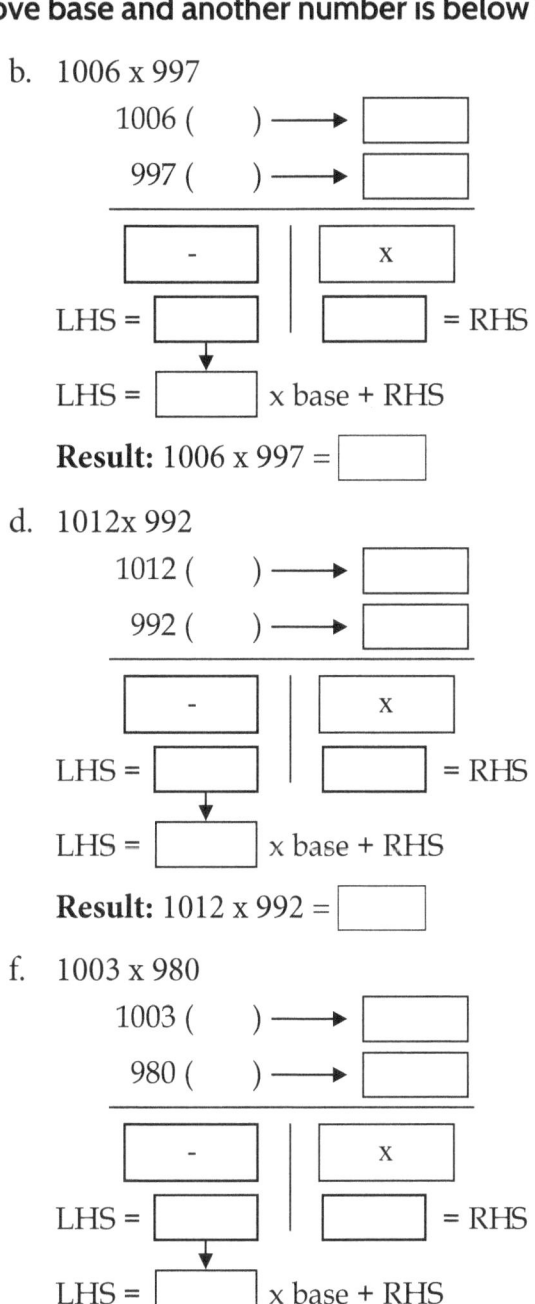

c. 11 x 8

d. 1012x 992

e. 103 x 88

f. 1003 x 980

II. Solve by base method:

a. 105 x 95   b. 12 x 7   c. 102 x 89

d. 1006 x 990   e. 1018 x 999   f. 102 x 80

g. 109 x 90

EXERCISE 3.4

I. Solve by base method: (Numbers with different base)

a. 56 x 880

56 x (   ) = _____

560 (   ) ⟶ ☐

880 (   ) ⟶ ☐

| ☐ - | ☐ x |

LHS = ☐ ‖ ☐ = RHS

Carry Over → ☐

LHS = ☐ + ☐ = ☐ /10

Result: 56 x 880 = ☐

b. 6 x 80

6 x (   ) = _____

60 (   ) ⟶ ☐

80 (   ) ⟶ ☐

| ☐ - | ☐ x |

LHS = ☐ ‖ ☐ = RHS

Carry Over → ☐

LHS = ☐ + ☐ = ☐ /10

Result: 6 x 80 = ☐

c. 32 x 360

32 x (   ) = _____

320 (   ) ⟶ ☐

360 (   ) ⟶ ☐

| ☐ - | ☐ x |

LHS = ☐ ‖ ☐ = RHS

Carry Over → ☐

LHS = ☐ + ☐ = ☐ /10

Result: 32 x 360 = ☐

d. 80 x 998

80 x (   ) = _____

800 (   ) ⟶ ☐

998 (   ) ⟶ ☐

| ☐ - | ☐ x |

LHS = ☐ ‖ ☐ = RHS

Carry Over → ☐

LHS = ☐ + ☐ = ☐ /10

Result: 80 x 998 = ☐

e. 28 x 420

28 x ( ) = _____

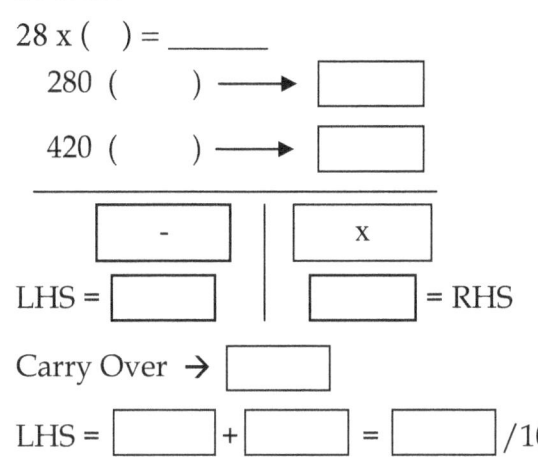

280 ( ) ⟶ ☐

420 ( ) ⟶ ☐

| - | | x |

LHS = ☐ | ☐ = RHS

Carry Over → ☐

LHS = ☐ + ☐ = ☐ /10

Result: 28 x 420 = ☐

f. 77 x 890

77 x ( ) = _____

770 ( ) ⟶ ☐

890 ( ) ⟶ ☐

| - | | x |

LHS = ☐ | ☐ = RHS

Carry Over → ☐

LHS = ☐ + ☐ = ☐ /10

Result: 77 x 890 = ☐

II. Solve by base method:

a. 90 x 980

b. 12 x 180

c. 68 x 780

d. 83 x 999

e. 22 x 450

f. 67 x 960

g. 11 x 130

EXERCISE 3.5

I. Solve the following (With higher number of 9's):

a. 632×999999

LHS = _____

RHS = _____

Result: $632 \times 999999 =$ ☐

b. 854×9999

LHS = _____

RHS = _____

Result: $854 \times 9999 =$ ☐

c. 45×9999

LHS = _____

RHS = _____

Result: $45 \times 9999 =$ ☐

d. 813×99999

LHS = _____

RHS = _____

Result: $813 \times 99999 =$ ☐

e. 86×9999

LHS = _____

RHS = _____

Result: $86 \times 9999 =$ ☐

f. 650×9999

LHS = _____

RHS = _____

Result: $650 \times 9999 =$ ☐

II.   **Solve the following:**

a. 234 × 99999

b. 12 × 999999

c. 736 × 9999

d. 654 × 99999

e. 6213 × 999999

f. 826 × 99999

EXERCISE 3.6

1. In a room, there is place for 70 tables. There are 102 rooms. How many tables can fill all the 102 rooms.

2. 116 tiles are required to cover a room. If there are 93 rooms, how many tiles are required to fill them?

3. An insect has 6 legs. How many legs do 12 insects have?

4. An electronic device takes 6 hours to charge. How many hours are required to charge 85 devices one after the other?

5. There are 106 desktop computers. Each requires 9 wire connections. How many wires are required?

6. Toffees come in packs of 8. How many toffees are there in 94 packs?

FAMOUS MATHEMATICIANS – EMMY NOETHER

	**Born:** 23 March 1882
	Died: 14 April 1935
	Nationality: German
	Famous For: Abstract algebra & Theoretical physics. Noether's work in abstract algebra and topology was influential in mathematics, which allowed her to approach problems of mathematics in fresh and original ways. As one of the leading mathematicians of her time, she developed the theories of rings, fields, and algebras. In physics, Noether's theorem explains the connection between symmetry and conservation laws.

CHAPTER 4

- MULTIPLICATION OF NUMBERS WITH DIFFERENT BASE (RECAP)

- MULTIPLICATION WHEN A BASE IS NOT POWER OF 10.

- SERIES OF 9'S

  ❖ MULTIPLICATION OF NUMBERS WITH LOWER NUMBER OF 9'S

MULTIPLICATION OF NUMBERS WITH DIFFERENT BASE (RECAP)

In multiplication of numbers with different bases means one may be closer to the base 10 and the other may be closer to the base 100 or base 1000 etc.

Consider an example, 80 x 981

In this example the multiplicand, 80 is closer to the base 100 and the multiplier, 981 is closer to the base 1000. Now let us discuss the above example.

Step 1: Here, the multiplicand is 80 and is closer to base 100, multiplier is 981 is closer to the base 1000. This can be represented as below:

$$80 \ (100) \longrightarrow -20$$
$$981(1000) \longrightarrow -19$$

Here, both the numbers are not with the same base. So, multiply the number with lower base with power of 10's so that both numbers will have same base. So, 80 x 10= 800.

Now, both the multiplier 800 and the multiplicand 981 are closer to the same base i.e. 1000. This can be represented as:

$$80 \ (100) \times 10 = 800 \ (1000)$$

$$981(1000) = 981(1000)$$

Step 2: Now, the multiplication is done by Nikhilam Sutra (Base method) discussed earlier. Find the differences. Here we must find the differences of the numbers from their respective base numbers. This can be represented as:

$$800 \ (1000) \longrightarrow -200$$
$$981 \ (1000) \longrightarrow -19$$

Step 3: Now, the answer is divided into 2 parts as LHS and RHS.

$$
\begin{array}{r}
800 \ (1000) \longrightarrow -200 \\
\times \ 981 \ (1000) \longrightarrow -19 \\
\hline
\text{LHS} \quad | \quad \text{RHS} \\
\hline
\end{array}
$$

Step 4: To find the RHS we multiply the differences on the RHS i.e., $200 \times 19 = 3800$. There must be only 3 digits in the RHS because the base (i.e. 1000) has 3 zeros. Thus, we carryover the 4th digit of 3800 (i.e. 3).

$$
\begin{array}{r|l}
800 \ (1000) \longrightarrow & -\ 200 \\
\times\ 981 \ (1000) \longrightarrow & -\ 19 \\
\hline
\text{LHS} & 800 \\
\end{array}
$$

Step 5: To find the LHS do the cross subtraction i.e. $800 - 19 = 781$. Add the carryover 3 to this i.e. $781 + 3 = 784$ and write it in the LHS. Now we write all the digits together which gives the answer.

$$
\begin{array}{r|l}
800 \ (1000) \longrightarrow & -\ 200 \\
\times\ 981 \ (1000) \longrightarrow & -\ 19 \\
\hline
784 & 800 \\
\end{array}
$$

So, $800 \times 981 = 784800$

Step 6: Now we must divide the above answer by the power of 10 with which we multiplied the lowest number out of the two to make them of the same base. So, we divide the answer obtained in Step 5 by 10 i.e. $784800/10 = 78480$

Result: $80 \times 981 = 78480$.

Consider an example, 78 x 983

Step 1: Multiply the lower of the two number from the power of 10 so that both the numbers will be of the same base.

So, $78 \times 10 = 780$

Now, both the multiplier 780 and the multiplicand 981 are closer to the same base i.e. 1000.

Step 2: Now, the multiplication is done by Nikhilam sutra (Base method). Here the base is 1000.

Step 3: Find the differences.

i.e., $780 \ (1000) \quad = \ -220$

And $983 \ (1000) \quad = \ -17$

Step 4: Now, the answer is divided into 2 parts as LHS and RHS.

$$
\begin{array}{r|l}
780 \ (1000) \longrightarrow & -\ 220 \\
\times\ 983 \ (1000) \longrightarrow & -\ 17 \\
\hline
\text{LHS} & \text{RHS} \\
\end{array}
$$

Step 5: To find the RHS we multiply the differences on the RHS i.e., $220 \times 17 = 3740$. There must be only 3 digits in the RHS because the base (i.e. 1000) has 3 zeros. Thus, we carryover the 4th digit of 3740 (i.e. 3).

$$780 \; (1000) \longrightarrow -220$$
$$\times 983 \; (1000) \longrightarrow -17$$

LHS	740

Step 6: To find the LHS do the cross subtraction i.e. $780 - 17 = 763$. Add the carryover 3 to this i.e. $763 + 3 = 766$ and write it in the LHS.

$$780 \; (1000) \longrightarrow -220$$
$$\times 983 \; (1000) \longrightarrow -17$$

766	740

Now we write all the digits together which gives the answer.

$780 \times 983 = 766740$

Step 7: Now we must divide the above answer by the power of 10 with which we multiplied the lowest number out of the two to make them of the same base. So, we divide the answer obtained in Step 6 by 10 i.e. $766740/10 = 76674$.

Result: 78 x 983 = 76674.

MULTIPLICATION WHEN A BASE IS NOT POWER OF 10

When the numbers are not closer to the base, we subtract them from the nearest base. We will get numbers as big as the given numbers itself. Thus, we make minor changes to find the base which will give us smaller difference when subtracted from it.

FOR BASE 100:

Consider an example, 39 x 47

Step 1: First we must find the base. The numbers are near to 50. 50 can be written as 100/2. Let us consider 50 to be the base number here.

Step 2: Let us find differences now. i.e., $50 - 39 = 11$ and $50 - 47 = 3$

$$39 \; (50) \longrightarrow 11$$
$$\times 47 \; (50) \longrightarrow 03$$

LHS	RHS

Step 3: Now, follow the Nikhilam sutra for multiplication.

$$39 \; (50) \longrightarrow 11$$
$$\times 47 \; (50) \longrightarrow 03$$

$39 - 03 = 36$	$11 \times 03 = 33$

Writing the digits of LHS and RHS together we get 3633 but this is not the right answer as the base is not power of 10 (i.e. 10, 100, 1000, etc.)

Step 4: Since the base considered in Step 1 is 50 which is half of 100 we divide the number on LHS by 2

$$
\begin{array}{r}
39\ (50) \longrightarrow 11 \\
\times\ 47\ (50) \longrightarrow 03 \\
\hline
36\ /\ 2 = 18\ \bigm|\ 33
\end{array}
$$

Result: 39 × 47 = 1833.

Consider an example, 40 x 43

Step 1: First we find the base. The numbers are near to 50. 50 can be written as 100/2. Let us consider 50 to be the base number here.

Step 2: Find the differences of the given numbers from the base. i.e. 50 – 40 = 10 and 50 – 43 = 7

$$
\begin{array}{r}
40\ (50) \longrightarrow 10 \\
\times\ 43\ (50) \longrightarrow 07 \\
\hline
\bigm|
\end{array}
$$

Step 3: Now, follow the Nikhilam sutra for multiplication.

$$
\begin{array}{r}
40\ (50) \longrightarrow 10 \\
\times\ 43\ (50) \longrightarrow 07 \\
\hline
40 - 7 = \mathbf{33}\ \bigm|\ 10 \times 07 = \mathbf{70}
\end{array}
$$

Write the digits of LHS and RHS together we get 3370 but this is not the right answer as the base is not power of 10 (i.e. 10, 100, 1000, etc.)

Step 4: Since the base considered in Step 1 is 50 which is half of 100 we divide the number on LHS by 2

$$
\begin{array}{r}
40\ (50) \longrightarrow 10 \\
\times\ 43\ (50) \longrightarrow 07 \\
\hline
33/2 = \mathbf{16\frac{1}{2}}\ \bigm|\ 10 \times 07 = \mathbf{70}
\end{array}
$$

This is not the required answer since the LHS contains fraction. Some manipulation is done to eliminate the fractional part from the answer.

Step 5: Now, as we know that 16½ is fraction and the multiplication of two non-fractional number cannot be a fractional number therefore, we will multiply the fraction ½ by the actual base i.e., 100 to obtain a non-fractional LHS then add it to the RHS.

So, ½ x 100 = 50

$$
\begin{array}{r}
40\ (50) \longrightarrow 10 \\
\times\ 43\ (50) \longrightarrow 07 \\
\hline
16\left(\frac{1}{2} \times 100\right) = \mathbf{50}\ \bigm|\ 70
\end{array}
$$

Add it to RHS we get 50 + 70 = 120. Thus, the new RHS will be 120.

$$
\begin{array}{r}
40\ (50) \longrightarrow 10 \\
\times\ 43\ (50) \longrightarrow 07 \\
\hline
16\ \mid\ 70 + 50 = \mathbf{120}
\end{array}
$$

We know that the number of digits in the RHS should not exceed the number of zeros in the base (100) so we should carryover 1 to the LHS. The LHS contains 16 to which we add the carryover 1 and we get 17 i.e. 16 + 1 = 17.

$$
\begin{array}{r}
40\ (50) \longrightarrow 10 \\
\times\ 43\ (50) \longrightarrow 07 \\
\hline
16 + 1 = \mathbf{17}\ \mid\ 20 \qquad = 17 \mid 20
\end{array}
$$

Result: 40 × 43 = 1720.

FOR BASE 1000:

Consider an example, 482 x 490

Step 1: First we must find the base. The numbers are near to 500. 500 can be written as 1000/2. Let us consider 500 to be the base number here.

Step 2: Find the differences of the given numbers from the base. i.e.,

500 – 482 = 18 and 500 – 490 = 10

$$
\begin{array}{r}
482\ (500) \longrightarrow 18 \\
\times\ 490\ (500) \longrightarrow 10 \\
\hline
\text{LHS} \mid \text{RHS}
\end{array}
$$

Step 3: Now, follow the Nikhilam sutra for multiplication.

$$
\begin{array}{r}
482\ (500) \longrightarrow 18 \\
\times\ 490\ (500) \longrightarrow 10 \\
\hline
482 - 10 = \mathbf{472} \mid 18 \times 10 = \mathbf{180}
\end{array}
$$

Write the digits of LHS and RHS together we get 472180 but this is not the right answer as the base is not power of 10.

Step 4: Since the base considered in Step 1 is 500 which is half of 1000 we divide the number on LHS by 2

$$
\begin{array}{r}
482\ (500) \longrightarrow 18 \\
\times\ 490\ (500) \longrightarrow 10 \\
\hline
472/2 = \mathbf{236} \mid 180
\end{array}
$$

Result: 482 × 490 = 236180.

MULTIPLICATION OF NUMBERS WITH LOWER NUMBER OF 9'S

In this section, we will learn how to multiply series of 9 when the digits in the multiplier are lesser than the digits in the multiplicand.

Consider an example, 9764 × 99

Here the multiplier has 2 digits whereas the multiplicand has 4 digits.

Step 1: The multiplier 99 can be written as (100-1).

Step 2: So, the problem can be represented as 9764 x (100-1).

Step 3: Multiplying 9764 by 100 and -1 we get,

9764 x 100 = 976400,

9764 x (-1) = – 9764

Now adding both we get,
   976400
-   9764
   ─────
   966636

Result: 9764 × 99 = 966636.

EXERCISE 4.1

I.   **Solve the following (Numbers with different base):**

a. 69 × 789

   69 × ( )   =   _____
   690 ( ) ⟶ ┌─────┐
   × 789 ( ) ⟶ │   │
              └─────┘
   ───────────────
   ┌─────┐ │ ┌─────┐
   │   │ │ │   │
   └─────┘ │ └─────┘
    LHS   │   RHS

   Result: 69 × 789 = _____

b. 80 × 884

   80 × ( )   =   _____
   800 ( ) ⟶ ┌─────┐
   × 884 ( ) ⟶ │   │
              └─────┘
   ───────────────
   ┌─────┐ │ ┌─────┐
   │   │ │ │   │
   └─────┘ │ └─────┘
    LHS   │   RHS

   Result: 80 × 884 = _____

c. 75 × 981

   75 × ( )   =   _____
   750 ( ) ⟶ ┌─────┐
   × 981 ( ) ⟶ │   │
              └─────┘
   ───────────────
   ┌─────┐ │ ┌─────┐
   │   │ │ │   │
   └─────┘ │ └─────┘
    LHS   │   RHS

   Result: 75 × 981 = _____

d. 89 × 973

   89 × ( )   =   _____
   890 ( ) ⟶ ┌─────┐
   × 973 ( ) ⟶ │   │
              └─────┘
   ───────────────
   ┌─────┐ │ ┌─────┐
   │   │ │ │   │
   └─────┘ │ └─────┘
    LHS   │   RHS

   Result: 89 × 973 = _____

e. 99 × 856

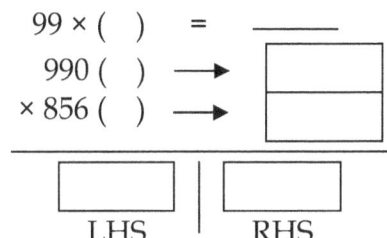

99 × ( ) = _____

990 ( ) ⟶

× 856 ( ) ⟶

LHS | RHS

Result: 99 × 856 = _____

f. 92 × 968

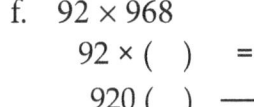

92 × ( ) = _____

920 ( ) ⟶

× 968 ( ) ⟶

LHS | RHS

Result: 92 × 968 = _____

II. Solve the following:

a. 89 × 991

b. 80 × 890

c. 63 × 974

d. 80 × 899

e. 78 × 962

f. 88 × 862

EXERCISE 4.2

I. Solve the following: (when a base is not power of 10) (for base 100)

a. 43 x 49

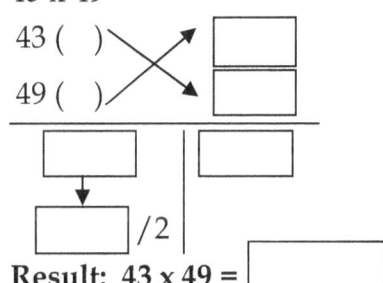

43 ( )

49 ( )

/2

Result: 43 x 49 = _____

b. 49 x 45

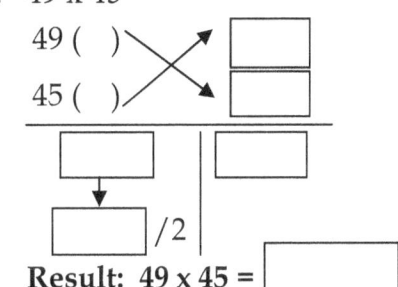

49 ( )

45 ( )

/2

Result: 49 x 45 = _____

c. 42 x 45

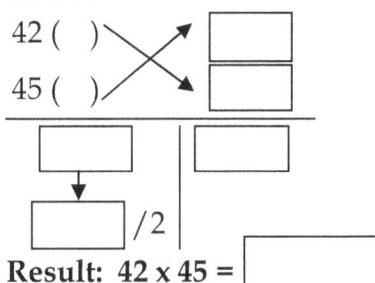

42 ( )

45 ( )

/2

Result: 42 x 45 = _____

d. 38 x 48

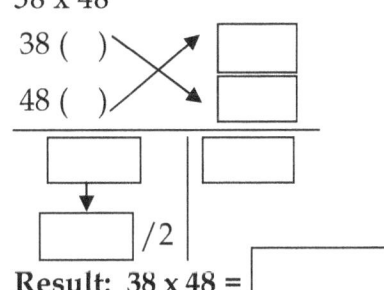

38 ( )

48 ( )

/2

Result: 38 x 48 = _____

e. 47 x 39

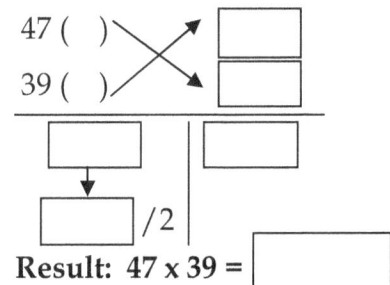

47 ( )

39 ( )

/2

Result: 47 x 39 = _____

f. 35 x 45

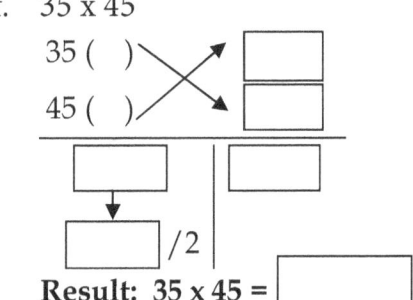

35 ( )

45 ( )

/2

Result: 35 x 45 = _____

II. Solve the following:

a. 41×31 b. 49×35 c. 50×38

d. 32×42 e. 43×39 f. 46×38

EXERCISE 4.3

I. Solve the following (when a base is not power of 10) (for base 1000)

a. 453 x 348

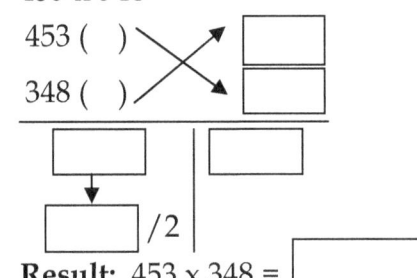

Result: 453 x 348 =

b. 400 x 345

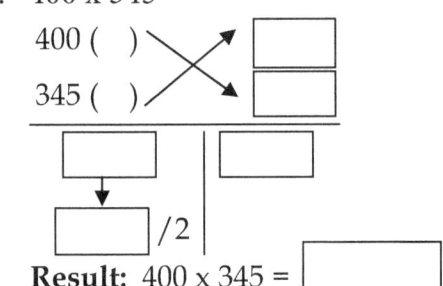

Result: 400 x 345 =

c. 401 x 449

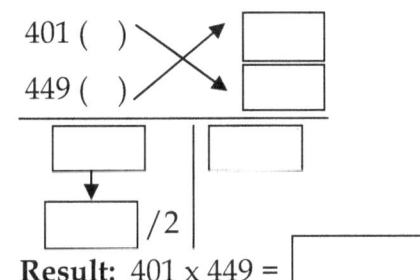

Result: 401 x 449 =

d. 358 x 345

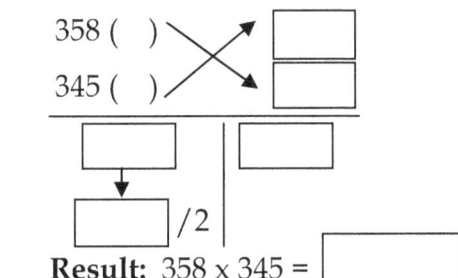

Result: 358 x 345 =

e. 498 x 399

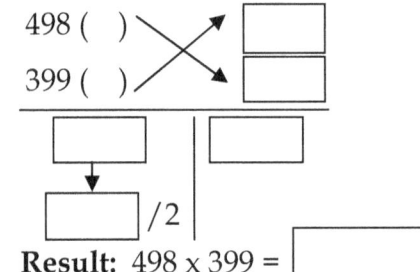

Result: 498 x 399 =

f. 429 x 389

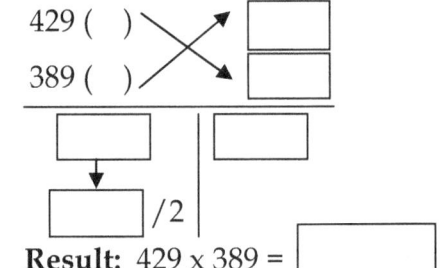

Result: 429 x 389 =

II. Solve the following:

a. 500×450 b. 480×399 c. 395×455

d. 408×453 e. 488×475 f. 469×498

EXERCISE 4.4

I. **Solve the following (With lower number of 9's):**

a. 863632 × 99999

LHS = _____

RHS = _____

Result: 863632 × 99999 = ☐

b. 6321854 × 9999

LHS = _____

RHS = _____

Result: 6321854 × 9999 = ☐

c. 45987 × 999

LHS = _____

RHS = _____

Result: 45987 × 999 = ☐

d. 185 × 99

LHS = _____

RHS = _____

Result: 185 × 99 = ☐

e. 7654386 × 9999

LHS = _____

RHS = _____

Result: 7654386 × 9999 = ☐

f. 527854 × 9999

LHS = _____

RHS = _____

Result: 527854 × 9999 = ☐

II. **Solve the following:**

a. 12 × 9

b. 657 × 99

c. 7654 × 999

d. 167 × 99

e. 27644 × 9999

f. 87512 × 999

EXERCISE 4.5

1. A pack has 8 lollipops. Arjun bought 92 packs. How many lollipops will Arjun get?
2. There are 9 racks. Each rack can hold 96 CD's. How many CD's can all the 9 racks hold?
3. On a tree, there are 6 branches. Each branch has 99 birds perched on it. How many birds are perched on the tree?
4. A man went to a shop to buy sweets. One pack contained 45 sweets. He bought 39 such packs. How many sweets did he buy?
5. A camera roll can capture 42 shots. There are 37 such cameras. How many shots can be taken?

FAMOUS MATHEMATICIANS – SATYENDRANATH BOSE

Born:	1 January 1894
Died:	4 February 1974
Nationality:	Indian

Famous For: Quantum Mechanics, Collaboration with Einstein.

Bose's works contributed to many concepts such as statistical mechanics, the electromagnetic properties of the ionosphere, the theories of X-ray crystallography and thermoluminescence, and unified field theory. His work also led Einstein to seek him out to work alongside him. Known for his collaboration with Albert Einstein. He is best known for his work on quantum mechanics in the early 1920s, providing the foundation for Bose–Einstein statistics and the theory of the Bose–Einstein condensate.

CHAPTER 5

- MULTIPLICATION BY 11 (2-DIGIT NUMBERS) WITHOUT CARRY OVER (RECAP)

- MULTIPLICATION WITH 11 (2-DIGIT NUMBERS) WITH CARRYOVER

- SERIES OF 1'S

  ❖ MULTIPLYING NUMBERS WITH EQUAL NUMBER OF 1's

  ❖ MULTIPLYING NUMBERS WITH EQUAL SERIES OF 1's

    ❑ MULTIPLYING WITH 3 ONE'S

    ❑ MULTIPLYING WITH 4 ONE'S

MULTIPLICATION BY 11 (RECAP)

MULTIPLICATION BY 11 (2-DIGIT NUMBERS) WITHOUT CARRYOVER

We have discussed about the multiplication of 11 in Chapter 5 (Level 1). Let us recall the method with the following examples.

Consider an example, 34 × 11.

The Result contains 3 parts among which 2 are the digits of the multiplicand itself i.e. we write 3 and 4 at both the ends.

$$3\,4 \times 1\,1$$

$$3 \mid \quad ? \quad \mid 4$$

At the center we write the sum of the digits of the multiplicand i.e. $3 + 4 = 7$.

$$3\,4 \times 1\,1$$

$$3 \mid \quad ? \quad \mid 4$$
$$= 3 \mid \quad 7 \quad \mid 4$$

Result: 34 x 11 = 374.

MULTIPLICATION BY 11 (2-DIGIT NUMBERS) WITH CARRYOVER

Now, we consider another example 79 × 11.

The Result contains 3 parts among which 2 are the digits of the multiplicand itself i.e. we write 7 and 9 at both the ends.

$$7\,9 \times 1\,1$$

$$7 \mid \quad ? \quad \mid 9$$

At the center we write the sum of the digits of the multiplicand i.e. $7 + 9 = 16$.

Since, there are 2 digits in the center we need to make it a single digit by carrying over the ten's place digit to the next digit (i.e. 7 + 1 = 8).

This can be represented as:

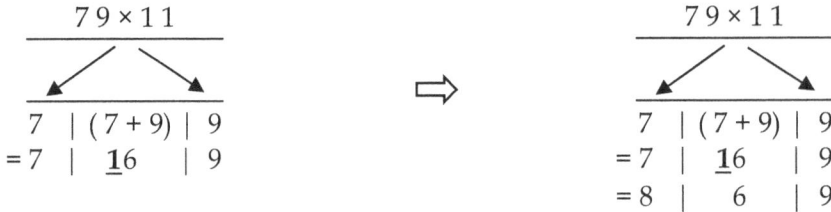

Result: 79 x 11 = 869.

SERIES OF 1'S

MULTIPLYING NUMBERS WITH EQUAL NUMBER OF 1'S

Consider the multiplication of two numbers where the multiplicand and multiplier are series of 1's (i.e. they contain only 1's) and the number of 1's are equal in them.

Like the multiplication of 11, we divide the Result into 3 parts as LHS, P and RHS (where, LHS – Left Hand Side, P – Pivotal and RHS – Right Hand Side).

First we count the number of 1's in the series of the multiplicand or multiplier. We write this number in the Pivotal part.

In the LHS and RHS we descend the numbers from the digit in the Pivotal part.

Now, let us learn with an example.

Consider, 111 × 111. We can write it as follows:

$$
\begin{array}{r}
1\,1\,1 \\
\times\,1\,1\,1 \\
\hline
\end{array}
$$

LHS	P	RHS

There are three 1's in the multiplicand or multiplier. So, in the Pivotal part of the Result we write 3.

$$
\begin{array}{r}
1\,1\,1 \\
\times\,1\,1\,1 \\
\hline
\end{array}
$$

LHS	3	RHS

Now, we write the descending numbers from the Pivotal part i.e. the numbers descending from 3. Thus, we have 2 and 1 in the LHS and RHS.

$$
\begin{array}{r}
1\,1\,1 \\
\times\,1\,1\,1 \\
\hline
\end{array}
$$

12	3	21

Result: 111 x 111 = 12321.

Let us consider another **example**, 11111 × 11111. We can write it as follows:

$$
\begin{array}{r}
1\,1\,1\,1\,1 \\
\times\,1\,1\,1\,1\,1 \\
\hline
\end{array}
$$

LHS	P	RHS

There are five 1's in the multiplicand or multiplier. So, in the Pivotal part of the Result we write 5.

$$
\begin{array}{r}
1\,1\,1\,1\,1 \\
\times\,1\,1\,1\,1\,1 \\
\hline
\end{array}
$$

LHS	5	RHS

Now, we write the descending numbers from the Pivotal part i.e. the numbers descending from 5. Thus, we have 4,3,2 and 1 in the LHS and RHS.

$$
\begin{array}{r}
1\,1\,1\,1\,1 \\
\times\,1\,1\,1\,1\,1 \\
\hline
\end{array}
$$

1234	5	4321

Result: 11111 x 11111 = 123454321.

MULTIPLYING NUMBERS WITH EQUAL SERIES OF 1'S

In this section, we see that the multiplier and multiplicand have the same number of digits and the multiplier has only one's (i.e. series of one's). We learn two cases in this section, one when the multiplier has 3 one's and the other when the multiplier has 4 one's.

First, we write the 1st digit of the multiplicand as it is and then add the digits in groups beginning from the right. The grouping of digits can be done upto the number of one's present in the multiplier. Let us have a detailed discussion with the examples.

> *NOTE:* The grouping of digits to find sum for the above method can be done only from **right to left**. No other pattern of grouping digits is followed.

MULTIPLYING WITH 3 ONE'S

WITHOUT CARRY OVER

The multiplier and multiplicand must have same number of digits and in this case the multiplier has 3 one's therefore the multiplicand also has 3 digits in it.

Consider an example, 342 × 111

$$
\underline{342 \times 111}
$$

Step 1: Write the last digit (i.e. the right most digit of the multiplicand) as it is in the right most of the Result. Here the multiplicand is 342 thus we write 2 as it is in the Result. This can be represented as:

$$342 \times 111$$
$$| \quad 2$$

Step 2: Find the sum of digits in groups beginning from right. First group 2 digits and next 3 digits. The group can be done to a maximum of 3 digits since 111 has 3 one's.

First, we group 2 and 4, their sum is 6 (i.e. 2 + 4 = 6). We write this in the Result before 2 i.e. we write it towards the left of 2. This can be represented as:

$$342 \times 111$$
$$| \quad 6 \,|\, 2$$

Step 3: Next we group 3 digits and find their sum i.e. 2, 4 and 3. (i.e. 2 + 4 + 3 = 9). We write this towards the left of 6. This can be represented as:

$$342 \times 111$$
$$| \quad 9 \,|\, 6 \,|\, 2$$

Step 4: Now all the possible groupings which begin with 2 have been done. We begin with the next digit i.e. 4. We group upto 3 digits since it's the maximum but here there are only two digits thus only groups of upto 2 can be done. Beginning with 4 the group of digits is 4 and 3 thus their sum is 7 (i.e. 4 + 3 = 7). We write it towards the left of 9 in the Result.

This can be represented as:

$$342 \times 111$$
$$7 \,|\, 9 \,|\, 6 \,|\, 2$$

Step 5: All the possible groupings beginning with 4 has been done. So, we begin with the next digit i.e. 3. It is a single digit or there are no numbers towards the left of 3 thus we write 3 in the Result towards the left of 7.

This can be represented as:

$$342 \times 111$$
$$3 \,|\, 7 \,|\, 9 \,|\, 6 \,|\, 2$$

Result: 342 x 111 = 37962.

Let us consider another example, 231 x 111

$$231 \times 111$$

Step 1: Write the last digit (i.e. the right most digit of the multiplicand) as it is in the right most of the Result. Here the multiplicand is 231 thus we write 1 as it is in the Result. This can be represented as:

$$231 \times 111$$
$$\boxed{1}$$

Step 2: Find the sum of digits in groups beginning from right. First group 2 digits and next 3 digits. The group can be done to a maximum of 3 digits since 111 has 3 one's.

First, we group 3 and 1, their sum is 4 (i.e. 3 + 1 = 4). We write this in the Result before 1 i.e. we write it towards the left of 1. This can be represented as

$$231 \times 111$$
$$\boxed{4 \mid 1}$$

Step 3: Next we group 3 digits and find their sum i.e. 2, 3 and 1. (i.e. 2 + 3 + 1 = 6). We write this towards the left of 4. This can be represented as:

$$231 \times 111$$
$$\boxed{6 \mid 4 \mid 1}$$

Step 4: Now, we begin with the next digit i.e. 3. We group upto 3 digits since it's the maximum but here there are only two digits thus only groups of upto 2 can be done. Beginning with 3 the group of digits is 3 and 2 thus their sum is 5 (i.e. 2 + 3 = 5). We write it towards the left of 6 in the Result.

This can be represented as:

$$231 \times 111$$
$$\boxed{5 \mid 6 \mid 4 \mid 1}$$

Step 5: All the possible groupings beginning with 3 has been done. So, we begin with the next digit i.e. 2. It is a single digit or there are no numbers towards the left of 2 thus we write 2 in the Result towards the left of 5.

This can be represented as:

$$231 \times 111$$
$$\boxed{2 \mid 5 \mid 6 \mid 4 \mid 1}$$

Result: 231 x 111 = 25641.

MULTIPLYING WITH 3 ONE'S (WITH CARRY OVER)

In the above example, we have seen the procedure for without carry over while finding sum i.e. the sum of digits after grouping did not exceed more than single digit. Suppose the sum exceeds single digit then we shall see how it must be carried over to obtain the Result.

Consider 576 × 111

Step 1: Write the last digit (i.e. the right most digit of the multiplicand) as it is in the right most of the Result. Here the multiplicand is 576 thus we write 6 as it is in the Result.

$$576 \times 111$$
$$6$$

Step 2: Find the sum of digits in groups beginning from right. First group 2 digits and next 3 digits. The group can be done to a maximum of 3 digits since 111 has 3 one's.

First, we group 6 and 7, their sum is 13 (i.e. 6 + 7 = 13). Here the sum contains 2 digits which cannot be written in the Result as it is. Thus, we write 3 in the Result towards the left of 6 and carryover 1.

$$576 \times 111$$
Carryover   1
$$3 \mid 6$$

Step 3: Next we group 3 digits and find their sum i.e. 6, 7 and 5. (i.e. 6 + 7 + 5 = 18). There is a carryover of 1 from the previous step which is added to this sum and we get 19 (i.e. 18 + 1 = 19). Here also the sum contains 2 digits thus we retain 9 in the Result and carryover 1. We write 9 towards the left of 3.

$$576 \times 111$$
Carryover   1
$$9 \mid 3 \mid 6$$

Step 4: Now all the possible groupings which begin with 6 have been done. We begin with the next digit i.e. 7. We group upto 3 digits since it's the maximum but here there are only two digits thus only groups of upto 2 can be done. Beginning with 7 the group of digits is 7 and 5 thus their sum is 12 (i.e. 7 + 5 = 12). There is a carryover of 1 from the previous step which is added to this sum and we get 13 (i.e. 12 + 1 = 13). Here also the sum contains 2 digits thus we retain 3 in the Result and carryover 1. We write 3 towards the left of 9 in the Result.

$$576 \times 111$$
Carryover 1
$$3 \mid 9 \mid 3 \mid 6$$

Step 5: All the possible groupings beginning with 7 has been done. So, we begin with the next digit i.e. 5. It is a single digit or there are no numbers towards the left of 5. Here there is a carryover of 1 from the previous step thus we add it to 5 which gives 6 (i.e. 5 + 1 = 6).

$$576 \times 111$$
$$6 \mid 3 \mid 9 \mid 3 \mid 6$$

Result: 576 x 111 = 63936.

MULTIPLYING WITH 4 ONE'S

The multiplier and multiplicand must have same number of digits and in this case the multiplier has 4 one's therefore the multiplicand also has 4 digits in it.

The grouping of digits can be done upto 4 digits since there are 4 one's in the multiplier.

Consider 4231 × 1111

Step 1: Write the last digit (i.e. the right most digit of the multiplicand) as it is in the right most of the Result. Here the multiplicand is 4231 thus we write 1 as it is in the Result.

$$4231 \times 1111$$
$$\boxed{1}$$

Step 2: Find the sum of digits in groups beginning from right. First group 2 digits then 3 digits later 4 digits. The group can be done to a maximum of 4 digits since 1111 has 4 one's.

First, we group 1 and 3, their sum is 4 (i.e. 1 + 3 = 4). We write 4 towards the left of 1.

$$4231 \times 1111$$
$$\boxed{4 \mid 6}$$

Step 3: Next we group 3 digits and find their sum i.e. 1, 3 and 2. (i.e. 1 + 3 + 2 = 6). We write 6 towards the left of 4.

$$4231 \times 1111$$
$$\boxed{6 \mid 4 \mid 1}$$

Step4: Now we group 4 digits and find their sum i.e. 1, 3, 2 and 4. (i.e. 1 + 3 + 2 + 4 = 10). Same as the previous case (i.e. Multiplying With 3 One's) we use carryover when the sum contains more than one digit. We write zero in the Result and carryover 1.

$$4231 \times 1111$$
Carryover   1
$$\boxed{0 \mid 6 \mid 4 \mid 1}$$

Step 5: All the possible groupings beginning with 1 has been done. So, we begin with the next digit i.e. 3. Groups of 4 digits cannot be made as there only 3 digits if we begin from 3 leaving 1. We add 3, 2 and 4 which gives 9. Since there is a carryover of 1 from the previous step we add it to the sum and we get 10. (i.e. 9 + 1 = 10). We write 0 towards the left of zero which is already there and carryover 1.

$$4231 \times 1111$$
Carryover   1
$$\boxed{0 \mid 0 \mid 6 \mid 4 \mid 1}$$

Step 6: Now we begin grouping with the next digit i.e. 2. The sum is 6 (i.e. 2 + 4 = 6). There is carryover of 1 from the previous step which is added to 6 and we get 7 (i.e. 6 + 1 = 7). Write 7 towards the left of 0.

$$4231 \times 1111$$

7	0	0	6	4	1

Step 7: Since grouping with 2 is done we begin with the next digit i.e. 4. As there are no digits to the left of 4 and no carryover from previous step we write it as it is in the Result.

$$4231 \times 1111$$

4	7	0	0	6	4	1

Result: 4231 x 1111 = 4700641.

EXERCISE 5.1

I. Multiplication by 11

a. 34×11 b. 47×11 c. 68×11

d. 26×11 e. 85×11 f. 93×11

EXERCISE 5.2

I.   Evaluate the following: (Multiplication with equal number of 1's)

a. 1111×1111

Number of 1's in the series (P) = _____

LHS	P	RHS

Result: 1111×1111 = [   ]

b. 111111×111111

Number of 1's in the series (P) = _____

LHS	P	RHS

Result: 111111×111111 = [   ]

c. 11111111×11111111

Number of 1's in the series (P) = _____

LHS	P	RHS

Result: 11111111×11111111= [   ]

d. 1111111×1111111

Number of 1's in the series (P) = _____

LHS	P	RHS

Result: 1111111×1111111 = [   ]

II.   Evaluate the following:

a. 111×111 b. 11111×11111 c. $111111111 \times 111111111$

EXERCISE 5.3

I.   Evaluate the following: (Multiplication with equal series of 1's – 3 one's Without carryover)

a. 241 × 111

241 × 111

⬚⬚⬚⬚⬚

Result: 241 × 111 = ⬚

b. 315 × 111

315 × 111

⬚⬚⬚⬚⬚

Result: 315 × 111 = ⬚

c. 152 × 111

152 × 111

⬚⬚⬚⬚⬚

Result: 152 × 111 = ⬚

d. 423 × 111

423 × 111

⬚⬚⬚⬚⬚

Result: 423 × 111 = ⬚

e. 305 × 111

305 × 111

⬚⬚⬚⬚⬚

Result: 305 × 111 = ⬚

f. 211 × 111

211 × 111

⬚⬚⬚⬚⬚

Result: 211 × 111 = ⬚

II.   Multiply below numbers with 111

a. 321

b. 412

c. 312

d. 413

e. 502

III.   Evaluate the following: (Multiplication with equal series of 1's – 3 one's With carryover)

a. 285 × 111

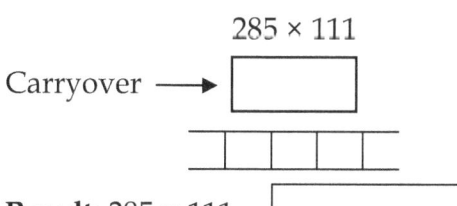

285 × 111

Carryover ⟶ ⬚

⬚⬚⬚⬚⬚

Result: 285 × 111 = ⬚

b. 634 × 111

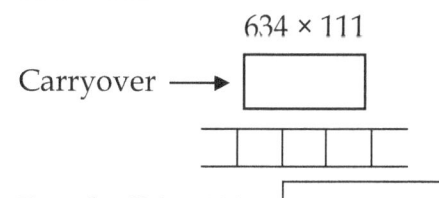

634 × 111

Carryover ⟶ ⬚

⬚⬚⬚⬚⬚

Result: 634 × 111 = ⬚

c. 847 × 111

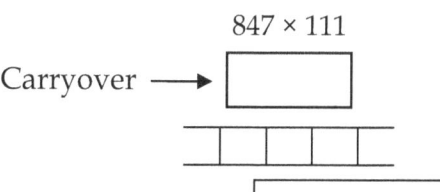

847 × 111

Carryover ⟶ ⬚

⬚⬚⬚⬚

Result: 847 × 111 = ⬚

d. 928 × 111

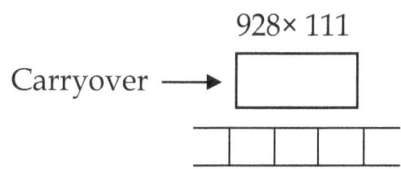

928× 111

Carryover ⟶ ⬚

⬚⬚⬚⬚

Result: 928× 111 = ⬚

IV. Multiply the below numbers with 111:

a. 134

b. 172

c. 375

d. 736

e. 263

f. 459

g. 547

h. 648

EXERCISE 5.4

I. Evaluate the following: (Multiplication with equal series of 1's – 4 one's)

a. 2463 × 1111

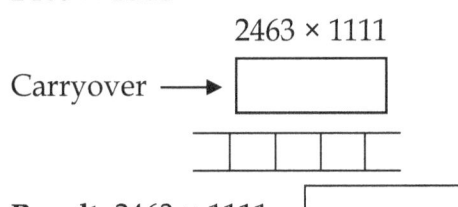

Result: 2463 × 1111 =

b. 1754 × 1111

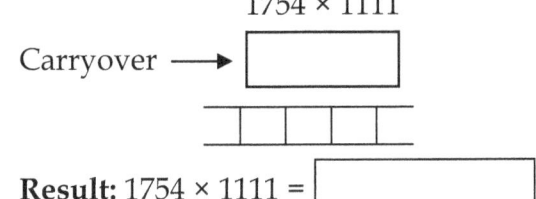

Result: 1754 × 1111 =

c. 5386 × 1111

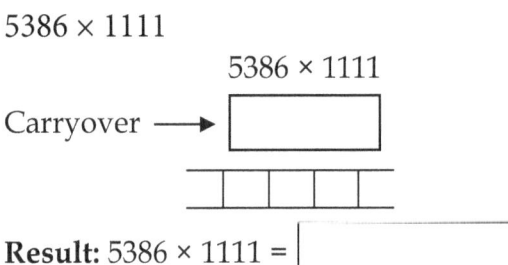

Result: 5386 × 1111 =

d. 7435 × 1111

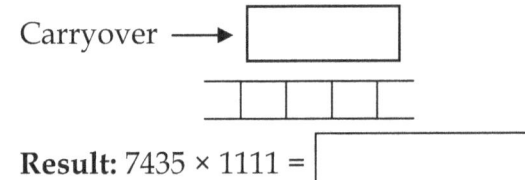

Result: 7435 × 1111 =

II. Evaluate the following:

a. 3627 × 1111

b. 4916 × 1111

c. 8542 × 1111

d. 7264 × 1111

e. 5478 × 1111

f. 9153 × 1111

EXERCISE 5.5

1. Betty went for a vacation to 11 places. She took 84 selfies in each location. How many photos does Betty have in all?

2. A teacher decided to take her students to a museum. The museum fare is Rs 175/- . There are 111 students in the class. How much money does the teacher have to pay to visit the museum?

3. Jia wanted to host a birthday party. She invited 235 guests. She wanted to give 111 tattoos to each of them. How many tattoos will Jia have to buy?

4. A girl went shopping and bought 111 shoes. Each of these shoes costed her Rs 111. How much money did the girl spend?

5. There are 1111 rooms. In each room, there are 1111 holes. Sandy wants to stick a photo on each hole in each of the 1111 rooms. How many photos does Sandy need, to do this?

FAMOUS MATHEMATICIANS – FIBONACCI

Born:	c. 1175
Died:	c. 1240-50
Nationality:	Republic of Pisa (Currently Italy)

Famous For: Fibonacci Series, Hindu – Arabic System.

Named 'the most talented Western Mathematicians of the middle ages. He is more correctly called Leonardo of Pisa or Leonardo Pisano. He grew up in North Africa, near the Mediterranean Coast from where he learnt how the traders did their arithmetic, thus realizing the advantage of Hindu – Arabic System. Thereby, replacing the Roman numeral system through his book Liber Abbaci.

He introduced the Fibonacci sequence which came from India, Arabia; which is famous till date. The sequence was named by a French Mathematician Edouard Lucas. This was derived when he tried to solve a riddle about reproduction in Rabbits.

CHAPTER 6

- DIVISIBILITY RULES OF 2 (RECAP)

- DIVISIBILITY RULES OF 3 (RECAP)

- DIVISIBILITY RULES OF 6

- DIVISIBILITY RULES OF 7

- DIVIDING A NUMBER BY 5 AND 9

- DIVISION

   ❖   USING NIKHILAM METHOD

DIVISION AND DIVISIBILITY RULES

We have already discussed the divisibility rules of 2 and 3 in Level – 1. Let us recall the rules of divisibility once again before going into divisibility of 6.

DIVISIBILITY RULE FOR 2

A number is divisible by 2 if its unit's place digit is even or zero.

DIVISIBILITY RULE FOR 3

A number is divisible by 3 if its digital root is divisible by 3.

DIVISIBILITY RULES FOR 6 AND 7

DIVISIBLE BY 6

A number is divisible by 6 if it is divisible by both 2 and 3.

Consider an example 138.

Check divisibility of 2: The unit's place digit is even (i.e. 8) so 138 is divisible by 2.

Check divisibility of 3: The digital root of 138 is 9 and 9 is divisible by 3 so 138 is also divisible by 3.

138 is divisible by 2 and 3 therefore, 138 is divisible by 6.

NOTE: If the number is divisible only by 2 or 3 then it is not divisible by 6.

Example: 256.

The unit's place digit is even (i.e. 6) so 256 is divisible by 2. The digital root of 256 is 7 and 7 is not divisible by 3 so 256 is not divisible by 3.

256 is divisible by 2 but not divisible by 3 therefore, 256 is not divisible by 6.

Example: 465.

The unit's place digit is odd (i.e. 5) so 465 is not divisible by 2. The digital root of 465 is 9 and 9 is divisible by 3 so 465 is divisible by 3.

465 is not divisible by 2 but divisible by 3 therefore, 465 is not divisible by 6.

DIVISIBLE BY 7

To check if a number is divisible by 7 we follow the steps below.

Step 1: Consider the unit's place digit of the given number and multiply it by 2.

Step 2: Now subtract the answer of Step 1 from the remaining digits in the given number.

Step 3: Check if the answer obtained in Step 2 is divisible by 7. If it is divisible by 7 then the given number is divisible by 7.

Consider an example 595.

Step 1: The unit's place digit is 5. Multiply 5 by 2 i.e. $5 \times 2 = 10$.

Step 2: The remaining digits in the given number are 59. So, subtract 10 from 59 i.e. $59 - 10 = 49$.

Step 3: 49 is divisible by 7 ($49 \div 7 = 7$, Remainder = 0). So, the given number i.e. 595 is divisible by 7.

The above steps can be represented as follows:

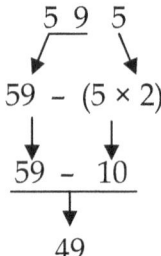

49 is divisible by 7.

Therefore, 595 is divisible by 7.

Let us consider another example 948

Step 1: The unit's place digit is 8. Multiply 8 by 2 i.e. $8 \times 2 = 16$.

Step 2: The remaining digits in the given number are 94. So, subtract 16 from 94 i.e. $94 - 16 = 78$.

Step 3: 78 is not divisible by 7 ($78 \div 7 = 11$, Remainder = 1). So, the given number i.e. 948 is not divisible by 7.

The above steps can be represented as follows:

```
 9 4 8
  ╱    ╲
94 – (8 × 2)
 ↓     ↓
94  –  16
 ────────
    78
```

78 is not divisible by 7.

Therefore, 948 is not divisible by 7.

Consider another example 1792.

Step 1: The unit's place digit is 2. Multiply 2 by 2 i.e. $2 \times 2 = 4$.

Step 2: The remaining digits in the given number are 179. So, subtract 4 from 179 i.e. $179 - 4 = 175$.

Step 3: Since 175 is a large number we cannot check if 175 is divisible by 7 orally. Thus, to find whether 175 is divisible by 7 we repeat the same process.

Step 4: The unit's place digit of 175 is 5. Multiply 5 by 2 i.e. $5 \times 2 = 10$.

Step 5: The remaining digits in the given number are 17. So, subtract 10 from 17 i.e. $17 - 10 = 7$.

Step 6: 7 is divisible by 7 ($7 \div 7 = 1$, Remainder $= 0$). So, the given number i.e. 1792 is divisible by 7.

The above steps can be represented as follows:

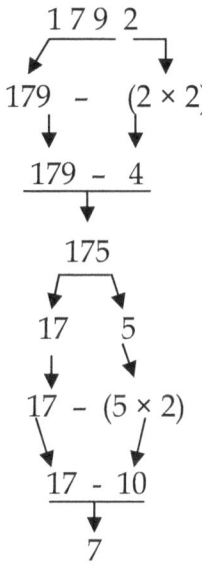

7 is divisible by 7.

Therefore, 1792 is divisible by 7.

DIVISION BY 5

We all know the conventional method to divide a number. In this section, we will learn an easy method to divide a number when the divisor is 5.

We shall see an example for conventional method of division say $17 \div 5$.

$$
\begin{array}{r}
5 \,)\, 17 \,(\, 3 \\
\underline{15} \\
\underline{02}
\end{array}
$$

We shall recall the terms involved in division.

$$\text{Divisor })\text{ Dividend }(\text{ Quotient}$$

$$\overline{\text{Remainder}}$$

In the example considered above, Dividend is 17, Divisor is 5, Quotient is 3 and Remainder is 2.

The easy method to divide when the divisor is 5 includes the following steps.

Step 1: First multiply the dividend by 2 so we get a new dividend.

Step 2: The answer obtained in Step 1 is divided by 10. Thus, we get the required answer.

Consider an example, 47 ÷ 5

Step 1: Multiply the dividend by 2. i.e. $47 \times 2 = 94$.

Step 2: Divide Step 1 answer by10. i.e. $94 \div 10 = 9.4$

The above steps can be represented as,

$$47 \times 2 = \frac{94}{10} = 9.4$$

Result: 47 ÷ 5 = 9.4.

The trick behind this method is we are multiplying both the dividend and divisor by 2 then divide the resulting numbers. As the divisor is 5 when we multiply it by 2 we get 10. It can be represented as follows,

$$\frac{47}{5} = \frac{47}{5} \times \frac{2}{2} = \frac{94}{10} = 9.4$$

DIVISION BY 9

We shall see an easy method to divide a number by 9 i.e. when the divisor is 9.

The steps included are given below.

Step 1: The first digit (the left most digit) of the number is written as it is in the answer part.

Step 2: Now add the 2nd digit of the given number to the digit written in the answer part of Step 1.

Step 3: The process is continued successively. The last digit (right most digit) in the answer part is the remainder. The rest of the digits form the quotient.

DIVISION BY 9 WITHOUT CARRYOVER IN THE QUOTIENT/REMAINDER PART

Consider an example, 53 ÷ 9.

Step 1: The given dividend is 53 and the left most digit is 5 which is written as it is in the answer part. It can be represented as,

$$9 \mid 5 \mid 3$$

Step 2: Now add the 2ⁿᵈ digit of the given number to the digit written in the answer part of Step 1. i.e. we add 3 to 5 as 3 is the 2ⁿᵈ digit of the dividend.

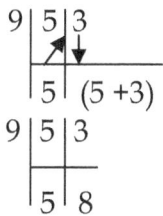

The above steps can be represented as,

The quotient is 5 and the remainder is 8.

Consider an example, 112 ÷ 9.

Step 1: The given dividend is 112 and the left most digit is 1 which is written as it is in the answer part. It can be represented as,

$$9 \mid 1 \; 1 \mid 2$$
$$\quad \; 1$$

Step 2: Now add the 2ⁿᵈ digit of the given number to the digit written in the answer part of Step 1. i.e. we add 1 to 1 as 1 is the 2ⁿᵈ digit of the dividend.

Step 3: We continue the process since there are digits towards right.

The above steps can be represented as,

The quotient is 12 and the remainder is 4.

DIVISION BY 9 WITH CARRYOVER IN THE QUOTIENT/REMAINDER PART

Consider an example, 137 ÷ 9

Step 1: The given dividend is 137 and the left most digit is 1 which is written as it is in the answer part. It can be represented as,

$$9\ |\ 1\ 3\ |\ 7$$
$$\downarrow$$
$$|\ 1$$

Step 2: Now add the 2nd digit of the given number to the digit written in the answer part of Step 1. i.e. we add 1 to 3 as 3 is the 2nd digit of the dividend.

$$9\ |\ 1\ 3\qquad |\ 7$$
$$\nearrow\!\downarrow$$
$$|\ 1\ (1+3)|$$

$$9\ |\ 1\ 3\ |\ 7$$
$$|\ 1\ 4|$$

Step 3: We continue the process since there are digits towards right.

$$9\ |\ 1\ 3\ |\ 7$$
$$\nearrow\!\downarrow$$
$$|\ 1\ 4\ |\ (7+4)$$

$$9\ |\ 1\ 3\ |\ 7$$
$$|\ 1\ 4\ |\ 11$$

The quotient is 14 and the remainder is 11.

Here, the remainder 11 is greater than the divisor. Thus, we continue the process of division on the remainder.

NOTE: One of the rules under division is that the remainder should not be greater than the divisor.

Step 4: Divide the remainder by the same method. The quotient obtained is added to the quotient of Step 3.

$$9\ |\ 1\ 3\ |\ 7$$
$$|\ 1\ 4\ |\ 11$$

$$9\ |\ 1\ |\ 1$$
$$\downarrow\ \nearrow\!\downarrow$$
$$|\ 1\ |\ (1+1)$$

$$9\ |\ 1\ |\ 1$$
$$|\ 1\ |\ 2$$

The quotient is 1 and the remainder is 2.

Add the quotient 1 obtained in this step to the quotient of Step 3 i.e. 1 + 14 = 15.

The remainder obtained in this step is the new remainder.

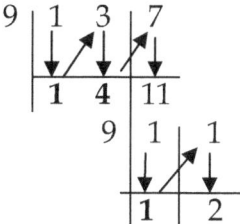

Thus, the quotient is 15 and the remainder is 2.

Consider an example, 283 ÷ 9

Step 1: The given dividend is 283 and the left most digit is 2 which is written as it is in the answer part. It can be represented as,

$$9 \; | \; \underset{\downarrow}{2} \; 8 \, | \, 3$$
$$\overline{\quad | \, 2 \quad | \quad}$$

Step 2: Now add the 2nd digit of the given number to the digit written in the answer part of Step 1. i.e. we add 2 to 8 as 8 is the 2nd digit of the dividend.

$$9 \, | \; 2 \quad \underset{\searrow}{8} \qquad | \; 3$$
$$\overline{\quad | \; 2 \; (8 + 2) |}$$
$$9 \, | \; 2 \quad 8 \quad | \; 3$$
$$\overline{\quad | \; 2 \quad 10 \; |}$$

Step 3: We continue the process since there are digits towards right.

$$9 \, | \; 2 \; 8 \; | \; \underset{\searrow}{3}$$
$$\overline{\quad | \; 2 \; 10 | \; (10 + 3)}$$
$$9 \, | \; 2 \; 8 \; | \; 3$$
$$\overline{\quad | \; 2 \; 10 | \; 13}$$

The quotient part contains 3-digits thus we must carryover 1 from 10. So, we add 1 to 2 to get 3 and 0 is at the same place. It can be represented as,

$$9 \; | \; 2 \; 8 | \; 3$$
$$\overline{\quad | \, 3 \; 0 | \, 13}$$

The normal carryover can be done only in the quotient part but not in the remainder part.

The quotient is 30 and the remainder is 13.

Here, the remainder 13 is greater than the divisor. Thus, we continue the process of division on the remainder.

Step 4: Divide the remainder by the same method. The quotient obtained is added to the quotient of Step 3.

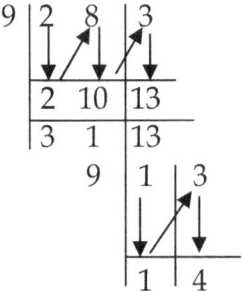

The quotient is 1 and the remainder is 4.

Add the quotient 1 obtained in this step to the quotient of Step 3 i.e. 1 + 30 = 31.

The remainder obtained in this step is the new remainder.

The above steps can be represented as,

Thus, the quotient is 31 and the remainder is 4.

DIVISION

CONVENTIONAL METHOD OF DIVISION

We all have studied the conventional method of division. Let us just recall the method. The usual representation is as given below.

Divisor) Dividend (Quotient

Remainder

The dividend and the divisor are written as above.

Step 1: Consider the left most digit of the dividend and divide it by the divisor. If the left most digit is smaller than the divisor then consider the 1ˢᵗ two digits and divide. The difference is found.

Step 2: Next to the difference write down the other digits which were not taken in Step 1. The other digits are brought down one by one. The process of division is continued.

Step 3: The division process is stopped once all the digits are brought down. The quotient and remainder are noted.

Suppose the remainder is greater than the divisor then the process is continued until the remainder is smaller than the divisor.

Consider an example 2435 ÷ 99

$$
\begin{array}{r}
99)\ 2435\ (24 \\
-\ 198\ \ \ \ \ \ \\
\hline
455\ \ \\
-\ 396\ \ \\
\hline
59\ \ \\
\hline
\end{array}
$$

Thus, the quotient is 24 and remainder is 59.

The conventional method is tedious especially when the numbers are large as in the considered example. We shall learn Nikhilam method of division which is easier even with larger numbers.

NIKHILAM METHOD OF DIVISION

In this section let us learn an easier method to do division especially when the numbers are large. The conventional method takes a long time in such cases whereas Nikhilam method takes just few minutes to solve it.

Let us discuss the Nikhilam method of division.

Step 1: Consider the given dividend and divisor. From the extreme right count the digits equal to the digits in the divisor and draw a slash. Thus, two parts are obtained in the process. The left part is called quotient block and the right part is called remainder block.

Step 2: Find the base closer to the divisor and find their difference.

Step 3: Multiply the difference with the first digit of the quotient part and write it below. Add the 1st digit of the dividend and the digits below it then write the sum in the quotient block.

Step 4: Multiply the difference with the second digit of the quotient part and write it below. Add the second digit of the dividend and the digits below it then write the sum in the quotient block.

Step 5: Now add the digits in the 3rd column and write the sum in the quotient block. The sum of the 3rd column is multiplied by the difference and written. It is continued until there are digits below the last digit of the dividend.

Step 6: If there are 2-digit in sum of the quotient block then carryover it to the previous digit (i.e. towards the left).

Step 7: The digits in the quotient block form the quotient and the digits in the remainder block form the remainder.

Consider an example, 476 ÷ 99.

Step 1: First separate the dividend into quotient and remainder blocks. 99 has 2 digits so the dividend must be separated after 2 digits from right. So, 76 will be in the remainder block and 4 will be in the quotient block.

It can be represented as,

$$
\begin{array}{c|c}
4 & 76 \\
\hline
 & \\
\end{array}
$$

Step 2: Find the base closer to the divisor. Here the divisor is 99, the closest base is 100 and the difference is 1. The difference is written as a 3-digit number since the base is 3-digit number (It is for calculation purpose only).

$$99\ (100)\quad 001$$

NOTE: The digits in the difference (i.e. between the base and divisor) must be equal to the digits in the base. This makes the calculation easier without confusion.

Step 3: Write the difference towards the left of the dividend and multiply with the 1st digit in the quotient block.

$$
\begin{array}{c|c|c}
001 & 4 & 7\ 6 \\
 & 0 & 0\ 4 \\
\hline
 & 4 & 7\ 10 \\
\end{array}
$$

Step 4: The remainder block contains 2 digits (i.e. 10) we must carryover it to 7.

Add 1 to 7 we get 8 (i.e. 1 + 7 = 8).

$$
\begin{array}{c|c|c}
001 & 4 & 7\ 6 \\
 & 0 & 0\ 4 \\
\hline
 & 4 & 8\ 0 \\
\end{array}
$$

The above steps can be represented as,

$$
\begin{array}{c|c|c}
001 & 4 & 7\ 6 \\
 & 0 & 0\ 4 \\
\hline
 & 4 & 7\ 10 \\
\hline
 & 4 & 8\ 0 \\
\end{array}
$$

The quotient is 4 and the remainder is 80.

Consider an example, 9987 ÷ 97

Step 1: First separate the dividend into quotient and remainder blocks. 97 has 2 digits so the dividend must be separated after 2 digits from right. So, 87 will be in the remainder block and 99 will be in the quotient block.

It can be represented as,

$$\begin{array}{c|c} 99 & 87 \\ \hline & \\ \end{array}$$

Step 2: Find the base closer to the divisor. Here the divisor is 97, the closest base is 100 and the difference is 3. The difference is written as a 3-digit number since the base is 3-digit number (It is for calculation purpose only).

$$97\ (100)\quad 003$$

Step 3: Write the difference towards the left of the dividend and multiply with the 1ˢᵗ digit in the quotient block.

$$\begin{array}{c|cc|cc} 003 & 9\ 9 & & 8\ 7 \\ & 0\ 2 & & 7 \\ & 0 & & 3\ 3 \\ \hline & 9\ 11 & & 18\ 10 \end{array}$$

Step 4: The remainder block contains 2 digits in both the sums so we must carryover it.

$$\begin{array}{c|cc|cc} 003 & 9\ 9 & & 8\ 7 \\ & 0\ 2 & & 7 \\ & 0 & & 3\ 3 \\ \hline & 10\ 1 & & 19\ 0 \end{array}$$

The remainder is 190 which is greater than the divisor so we again continue the division process until we get remainder as zero or smaller than divisor.

Step 5: Divide 190 using the same process explained in the above steps.

$$\begin{array}{c|c|cc} 003 & 1 & & 9\ 0 \\ & 0 & & 0\ 3 \\ \hline & 1 & & 9\ 3 \end{array}$$

The quotient is 1 and the remainder is 93. Add the quotient to the quotient of Step 4 to get the required result.

i.e. 101 + 1 = 102.

The above steps can be represented as,

$$\begin{array}{c|cc|cc} 003 & 9\ 9 & & 8\ 7 \\ & 0\ 2 & & 7 \\ & 0 & & 3\ 3 \\ \hline & 9\ 11 & & 18\ 10 \\ \hline & 10\ 1 & & 19\ 0 \\ & 003\quad 1 & & 9\ 0 \\ & 0 & & 0\ 3 \\ \hline & 1 & & 9\ 3 \end{array}$$

The quotient is 102 and the remainder is 93.

EXERCISE 6.1

I. Find whether the given numbers are divisible by 2:

a. 783

Unit's place digit is _____, which is even/odd number.

Result: _____ is divisible/not divisible by 2.

b. 372

Unit's place digit is _____, which is even/odd number.

Result: _____ is divisible/not divisible by 2.

c. 122358

Unit's place digit is _____, which is even/odd number.

Result: _____ is divisible/not divisible by 2.

d. 833779

Unit's place digit is _____, which is even/odd number.

Result: _____ is divisible/not divisible by 2.

e. 654893

Unit's place digit is _____, which is even/odd number.

Result: _____ is divisible/not divisible by 2.

f. 4785216

Unit's place digit is _____, which is even/odd number.

Result: _____ is divisible/not divisible by 2.

II. Find whether the given numbers are divisible by 3:

a. 846

Digital root is _____, which is divisible/ not divisible by 3.

Result: _____ is divisible/not divisible by 3.

b. 452

Digital root is _____, which is divisible/ not divisible by 3.

Result: _____ is divisible/not divisible by 3.

c. 1964

Digital root is _____, which is divisible/ not divisible by 3.

Result: _____ is divisible/not divisible by 3.

d. 2548

Digital root is _____, which is divisible/ not divisible by 3.

Result: _____ is divisible/not divisible by 3.

e. 46657

Digital root is _____, which is divisible/ not divisible by 3.

Result: _____ is divisible/not divisible by 3.

f. 65523

Digital root is _____, which is divisible/ not divisible by 3.

Result: _____ is divisible/not divisible by 3.

EXERCISE 6.2

I.   Find whether the given number are divisible by 6:

a. 186

_____ is divisible/not divisible by 2.

Digital root of 186 is _____ so, 186 is divisible/not divisible by 3.

Result: _____ is divisible/not divisible by 6.

b. 537

_____ is divisible/not divisible by 2.

Digital root of 537 is _____ so, 537 is divisible/not divisible by 3.

Result: _____ is divisible/not divisible by 6.

c. 4987

_____ is divisible/not divisible by 2.

Digital root of 4987 is _____ so, 4987 is divisible/not divisible by 3.

Result: _____ is divisible/not divisible by 6.

d. 7342

_____ is divisible/not divisible by 2.

Digital root of 7342 is _____ so, 7342 is divisible/not divisible by 3.

Result: _____ is divisible/not divisible by 6.

e. 21168

_____ is divisible/not divisible by 2.

Digital root of 21168 is _____ so, 21168 is divisible/not divisible by 3.

Result: _____ is divisible/not divisible by 6.

f. 59184

_____ is divisible/not divisible by 2.

Digital root of 59184 is _____ so, 59184 is divisible/not divisible by 3.

Result: _____ is divisible/not divisible by 6.

g. 80820

_____ is divisible/not divisible by 2.

Digital root of 80820 is _____ so, 80820 is divisible/not divisible by 3.

Result: _____ is divisible/not divisible by 6.

h. 95783

_____ is divisible/not divisible by 2.

Digital root of 95783 is _____ so, 95783 is divisible/not divisible by 3.

Result: _____ is divisible/not divisible by 6.

II.   Find whether the given numbers are divisible by 6:

a. 2954

b. 7176

c. 9792

d. 5203

e. 34675

f. 56814

g. 64137

h. 21288

EXERCISE 6.3

I. Find whether the given numbers are divisible by 7:

a. 156

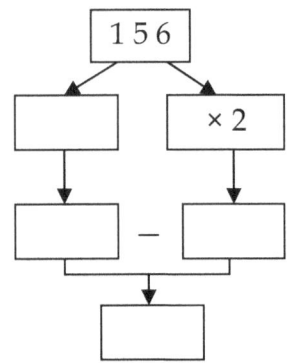

_____ is divisible/not divisible by 7.

Result: _____ is divisible/not divisible by 7.

b. 399

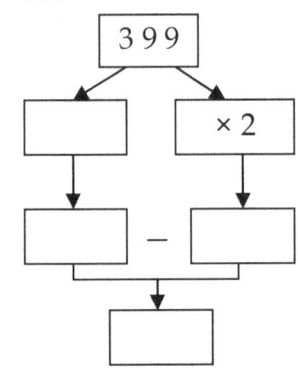

_____ is divisible/not divisible by 7.

Result: _____ is divisible/not divisible by 7.

c. 742

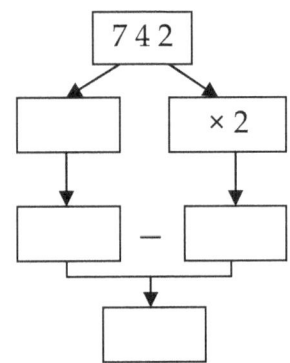

_____ is divisible/not divisible by 7.

Result: _____ is divisible/not divisible by 7.

d. 986

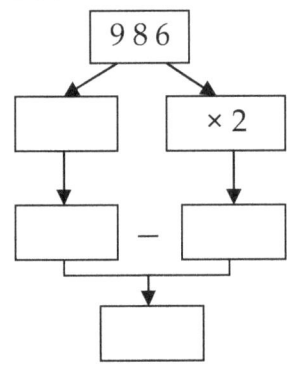

_____ is divisible/not divisible by 7.

Result: _____ is divisible/not divisible by 7.

e. 2453

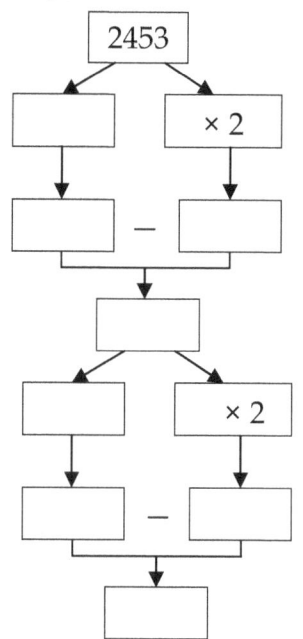

_____ is divisible/not divisible by 7.

Result: _____ is divisible/not divisible by 7.

f. 6055

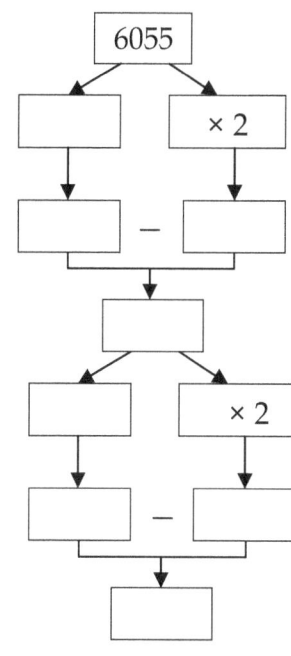

_____ is divisible/not divisible by 7.

Result: _____ is divisible/not divisible by 7.

II. **Find whether the given numbers are divisible by 7:**

a. 3689

b. 3572

c. 7990

d. 8610

e. 12047

f. 27744

g. 67285

h. 66801

EXERCISE 6.4

I. **Divide the given numbers by 5:**

a. 38

$$\frac{\_\_\_ \times 2 = \_\_\_}{10} = \boxed{}$$

Result: 38 ÷ 5 = $\boxed{}$

b. 57

$$\frac{\_\_\_ \times 2 = \_\_\_}{10} = \boxed{}$$

Result: 57 ÷ 5 = $\boxed{}$

c. 72

$$\frac{\_\_\_ \times 2 = \_\_\_}{10} = \boxed{}$$

Result: 72 ÷ 5 = $\boxed{}$

d. 94

$$\frac{\_\_\_ \times 2 = \_\_\_}{10} = \boxed{}$$

Result: 94 ÷ 5 = $\boxed{}$

e. 122

$$\frac{\_\_\_ \times 2 = \_\_\_}{10} = \boxed{}$$

Result: 122 ÷ 5 = $\boxed{}$

f. 173

$$\frac{\_\_\_ \times 2 = \_\_\_}{10} = \boxed{}$$

Result: 173 ÷ 5 = $\boxed{}$

g. 216

___ × 2 = ___ = ⬚
   10

Result: 216 ÷ 5 = ⬚

h. 258

___ × 2 = ___ = ⬚
   10

Result: 258 ÷ 5 = ⬚

II. **Divide the given numbers by 5:**

a. 49

b. 66

c. 81

d. 99

e. 117

f. 134

g. 197

h. 243

EXERCISE 6.5

I. **Divide the given numbers by 9 (Without carryover in the quotient /remainder part):**

a. 43

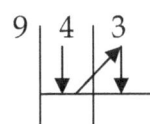

Result: The quotient is _____ and the remainder is _____.

b. 17

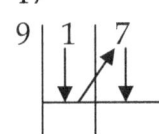

Result: The quotient is _____ and the remainder is _____.

c. 32

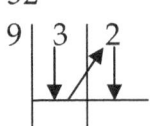

Result: The quotient is _____ and the remainder is _____.

d. 51

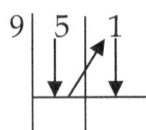

Result: The quotient is _____ and the remainder is _____.

e. 114

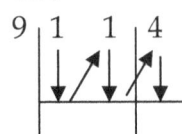

Result: The quotient is _____ and the remainder is _____.

f. 222

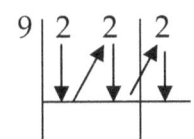

Result: The quotient is _____ and the remainder is _____.

II.   **Divide the given numbers by 9 (Without carryover in the quotient /remainder part):**

a. 33

b. 26

c. 71

d. 44

e. 123

f. 142

g. 214

h. 233

EXERCISE 6.6

I.   **Divide the given numbers by 9 (With carryover in the quotient /remainder part):**

a. 38

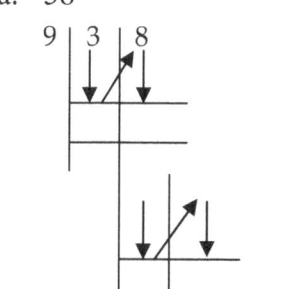

Result: Quotient =

Remainder =

b. 79

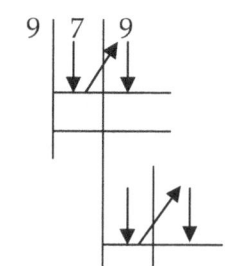

Result: Quotient =

Remainder =

c. 184

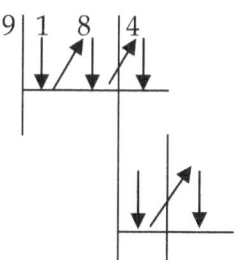

Result: Quotient =

Remainder =

d. 239

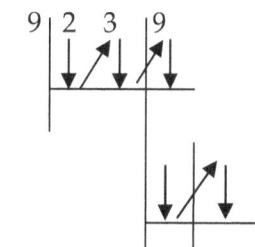

Result: Quotient =

Remainder =

e. 386

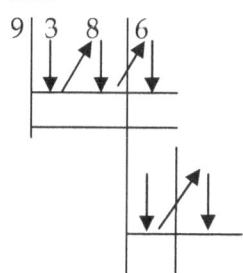

Result: Quotient =

Remainder =

f. 596

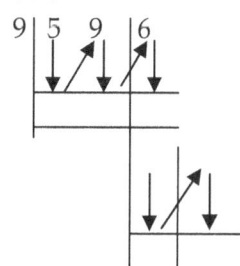

Result: Quotient =

Remainder =

g. 735

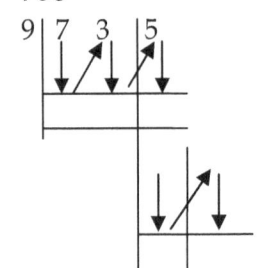

Result: Quotient = ☐

Remainder = ☐

h. 462

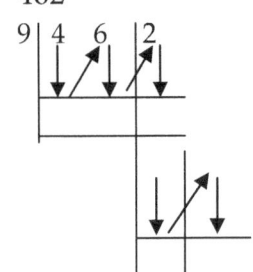

Result: Quotient = ☐

Remainder = ☐

II.   Divide the given numbers by 9 (With carryover in the quotient /remainder part):

a. 68

b. 47

c. 95

d. 126

e. 177

f. 649

g. 824

h. 923

EXERCISE 6.7

I.   Divide the given numbers by Nikhilam method:

a. 586 ÷ 97

97 (   ) ⟶ ☐

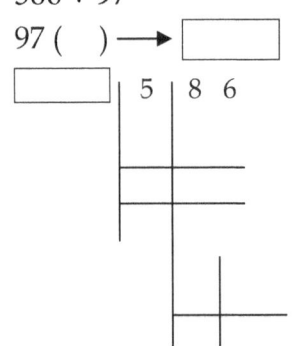

Result: Quotient = ☐

Remainder = ☐

b. 375 ÷ 89

89 (   ) ⟶ ☐

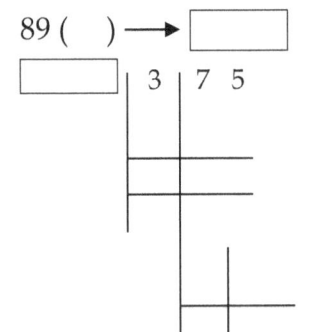

Result: Quotient = ☐

Remainder = ☐

c. 2345 ÷ 98

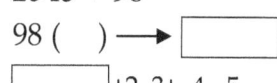

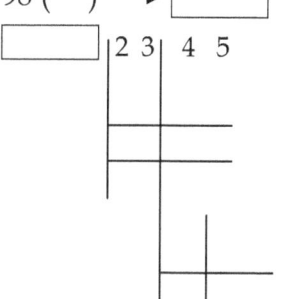

Result: Quotient = [   ]

Remainder = [   ]

d. 4012 ÷ 86

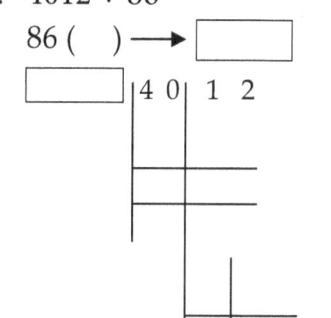

Result: Quotient = [   ]

Remainder = [   ]

e. 156 ÷ 87

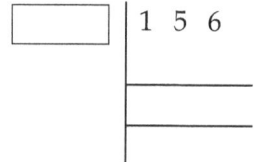

Result: Quotient = [   ]

Remainder = [   ]

f. 1230 ÷ 95

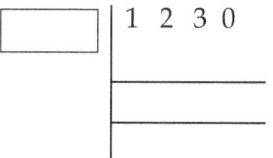

Result: Quotient = [   ]

Remainder = [   ]

g. 8367 ÷ 96

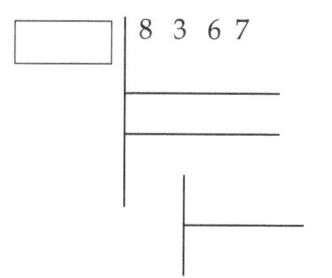

Result: Quotient = [   ]

Remainder = [   ]

h. 6473 ÷ 99

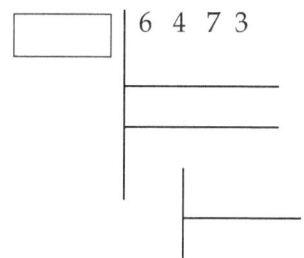

Result: Quotient = [   ]

Remainder = [   ]

II. Divide the given numbers by Nikhilam method:

a. 225 ÷ 85

b. 786 ÷ 83

c. 1111 ÷ 89

d. 2345 ÷ 87

e. 345 ÷ 93

f. 632 ÷ 98

g. 1234 ÷ 92

h. 3344 ÷ 99

FAMOUS MATHEMATICIANS – DANIEL BERNOULLI

Born:	8 February 1700
Died:	17 March 1782
Nationality:	Swiss

Famous For: He is particularly remembered for his applications of mathematics to mechanics, especially fluid mechanics, and for his pioneering work in probability and statistics. His name is commemorated in the Bernoulli's principle. a particular example of the conservation of energy, which describes the mathematics of the mechanism underlying the operation of two important technologies of the 20th century: the carburetor and the airplane wing.

ANSWERS

EXERCISE 1.1

I. a. 165
   b. 276
   c. 110
   d. 240
   e. 363
   f. 242
   g. 396
   h. 483

II. a. 253
   b. 484
   c. 168
   d. 294
   e. 682

EXERCISE 1.2

I. a. 26331
   b. 25953
   c. 33712
   d. 27951
   e. 63812
   f. 28820
   g. 32421
   h. 11312

II. a. 38841
   b. 65932
   c. 68420
   d. 22220
   e. 44440

EXERCISE 1.3

1. 23460
2. 11312
3. 24888

EXERCISE 2.1

I. a. 957
   b. 780
   c. 598
   d. 3510
   e. 1904
   f. 434
   g. 768
   h. 708

II. a. 528
   b. 2196
   c. 899
   d. 4104
   e. 703

EXERCISE 2.2

I. a. 175404
   b. 198501
   c. 676620
   d. 230594
   e. 24435
   f. 56784
   g. 35496
   h. 119341

II. a. 180348

   b. 307800

   c. 49410

   d. 485865

EXERCISE 2.3

I. a. 3366

   b. 764235

   c. 346653

   d. 123876

   e. 86331366

   f. 54344565

II. a. 144855

   b. 7653243123467568

   c. 38786121

   d. 1188

   e. 75312468

   f. 133866

EXERCISE 2.4

1. 24603

2. 43524

3. 22878

EXERCISE 3.1

I. a. 100, below

   b. 100, above

   c. 100above

   d. 1000, above

   e. 10000, below

   f. 1000, below

   g. 10, above

   h. 10, below

II. a. 9212

   b. 11024

   c. 168

   d. 984048

   e. 11016

   f. 991008

III. a. 11448

   b. 9009

   c. 10605

   d. 893606

   e. 988020

   f. 9603

   g. 987030

EXERCISE 3.2

I. a. 783200

   b. 7020

   c. 111554

   d. 709600

   e. 1039378

   f. 664440

II. a. 572000

   b. 30240

   c. 538780

   d. 778926

   e. 77000

   f. 864000

   g. 18000

EXERCISE 3.3

I. a. 10098

   b. 1002982

   c. 88

   d. 1003904

   e. 9064

   f. 982940

II. a. 9975

   b. 84

   c. 9078

   d. 995940

   e. 1016982

f. 8160

g. 9810

4. 510

5. 954

6. 752

EXERCISE 3.4

I.   a. 48280

b. 480

c. 11520

d. 79840

e. 11760

f. 68530

II. a. 88200

b. 2160

c. 53040

d. 82917

e. 9900

f. 64320

g. 1430

EXERCISE 3.5

I.   a. 631999368

b. 8539146

c. 449955

d. 81299187

e. 859914

f. 6499350

II. a. 23399766

b. 11999988

c. 7359264

d. 65399346

e. 6212993787

f. 82599174

EXERCISE 3.6

1. 7140

2. 10788

3. 72

EXERCISE 4.1

I.   a. 54441

b. 70720

c. 73575

d. 86597

e. 84744

f. 89056

II. a. 88199

b. 71200

c. 61362

d. 71920

e. 75036

f. 75856

EXERCISE 4.2

I.   a. 2107

b. 2205

c. 1890

d. 1824

e. 1833

f. 1575

II. a. 1271

b. 1715

c. 1900

d. 1344

e. 1677

f. 1748

EXERCISE 4.3

I.   a. 157644

b. 138000

c. 180049

d. 123510

e. 198702

f. 166881

II. a. 225000

b. 191520

c. 179725

d. 184824

e. 231800

f. 233562

EXERCISE 4.4

I. a. 86362336368

b. 63212218146

c. 45941013

d. 18315

e. 76536205614

f. 5278012146

II. a. 108

b. 65043

c. 7646346

d. 16533

e. 276412356

f. 87424488

EXERCISE 4.5

1. 736

2. 864

3. 594

4. 1755

5. 1554

EXERCISE 5.1

I. a. 374

b. 517

c. 748

d. 286

e. 935

f. 1023

EXERCISE 5.2

I. a. 1234321

b. 12345654321

c. 123456787654321

d. 1234567654321

II. a. 12321

b. 123454321

c. 12345678987654321

EXERCISE 5.3

I. a. 26751

b. 34965

c. 16872

d. 46953

e. 33855

f. 23421

II. a. 35631

b. 45732

c. 34632

d. 45843

e. 55722

III. a. 31635

b. 70374

c. 94017

d. 103008

IV. a. 14874

b. 19092

c. 41625

d. 81696

e. 29193

f. 50949

g. 60717

h. 71928

EXERCISE 5.4

I. a. 2736393

b. 1948694

c. 5983846

d. 8260285

II. a. 4029597

b. 5461676

c. 9490162

d. 8070304

e. 6086058

f. 10168983

EXERCISE 5.5

1. 924

2. 19425

3. 26085

4. 12321

5. 1234321

EXERCISE 6.1

I. a. Not divisible

b. Divisible

c. Divisible

d. Not divisible

e. Not divisible

f. Divisible

II. a. Divisible

b. Not divisible

c. Not divisible

d. Not divisible

e. Not divisible

f. Divisible

EXERCISE 6.2

I. a. Divisible

b. Not divisible

c. Not divisible

d. Not divisible

e. Divisible

f. Divisible

g. Divisible

h. Not divisible

II. a. Not divisible

b. Divisible

c. Divisible

d. Not divisible

e. Not divisible

f. Divisible

g. Not divisible

h. Divisible

EXERCISE 6.3

I. a. Not divisible

b. Divisible

c. Divisible

d. Not divisible

e. Not divisible

f. Divisible

II. a. Divisible

b. Not divisible

c. Not divisible

d. Divisible

e. Divisible

f. Not divisible

g. Not divisible

h. Divisible

EXERCISE 6.4

I. a. 7.6

b. 11.4

c. 14.4

d. 18.8

e. 24.4

f. 34.6

g. 43.2

h. 51.6

II. a. 9.8

b. 13.2

c. 16.2

d. 19.8

e. 23.4

f. 26.8

g. 39.4

h. 48.6

EXERCISE 6.5

I. a. Q-4, R-7

b. Q-1, R-8

c. Q-3, R-5

d. Q-5, R-6

e. Q-12, R-6

f. Q-24, R 6

II. a. Q-3, R-6

b. Q-2, R-8

c. Q-7, R-8

d. Q-4, R-8

e. Q-13, R-6

f. Q-15, R-7

g. Q-23, R-7

h. Q-25, R-8

EXERCISE 6.6

I. a. Q-4, R-2

b. Q-8, R-7

c. Q-20, R-4

d. Q-26, R-5

e. Q-42, R-8

f. Q-66, R-2

g. Q-81, R-6

h. Q-51, R-3

II. a. Q-7, R-5

b. Q-5, R-2

c. Q-10, R-5

d. Q-14, R-0

e. Q-19, R-6

f. Q-72, R-1

g. Q-91, R-5

h. Q-102, R-5

EXERCISE 6.7

I. a. Q-6, R-4

b. Q-4, R-19

c. Q-23, R-91

d. Q-46, R-56

e. Q-1, R-69

f. Q-12, R-90

g. Q-87, R-15

h. Q-65, R-38

II. a. Q-2, R-55

b. Q-9, R-39

c. Q-12, R-43

d. Q-26, R-83

e. Q-3, R-66

f. Q-6, R-44

g. Q-13, R-38

h. Q-33, R-77